FE ELECTRICAL AND COMPUTER PRACTICE PROBLEMS

for the Electrical and Computer Fundamentals of Engineering Exam

Michael R. Lindeburg, PE

PPI®
PPI2PASS.COM

Professional Publications, Inc. • Belmont, California

Report Errors and View Corrections for This Book

Everyone benefits when you report typos and other errata, comment on existing content, and suggest new content and other improvements. You will receive a response to your submission; other engineers will be able to update their books; and, PPI will be able to correct and reprint the content for future readers.

PPI provides two easy ways for you to make a contribution. If you are reviewing content on **feprep.com**, click on the "Report a Content Error" button in the Help and Support area of the footer. If you are using a printed copy of this book, go to **ppi2pass.com/errata**. To view confirmed errata, go to **ppi2pass.com/errata**.

FE ELECTRICAL AND COMPUTER PRACTICE PROBLEMS

Current release of this edition: 1

Release History

date	edition number	revision number	update
Apr 2017	1	1	New book.

Printed in the United States of America.

PPI
1250 Fifth Avenue, Belmont, CA 94002
(650) 593-9119
ppi2pass.com

ISBN: 978-1-59126-450-7

Library of Congress Control Number: 2017933813

F E D C B A

Topics

Topic I:	Mathematics
Topic II:	Probability and Statistics
Topic III:	Properties of Electrical Materials
Topic IV:	Engineering Sciences
Topic V:	Electromagnetics
Topic VI:	Circuit Analysis and Linear Systems
Topic VII:	Power
Topic VIII:	Electronics
Topic IX:	Control Systems
Topic X:	Communications and Signal Processing
Topic XI:	Computer Networks and Systems
Topic XII:	Digital Systems
Topic XIII:	Software Development
Topic XIV:	Engineering Economics
Topic XV:	Ethics and Professional Practice

Where do I find help solving these Practice Problems?

FE Electrical and Computer Practice Problems presents complete, step-by-step solutions for more than 450 problems to help you prepare for the Electrical and Computer FE exam. You can find all the background information, including charts and tables of data, that you need to solve these problems in the *FE Electrical and Computer Review Manual.*

The *FE Electrical and Computer Review Manual* may be obtained from PPI at **ppi2pass.com** or **feprep.com**, or from your favorite print book retailer.

Table of Contents

Preface

The purpose of this book is to prepare you for the National Council of Examiners for Engineering and Surveying (NCEES) Fundamentals of Engineering (FE) exam.

In 2014, the NCEES adopted revised specifications for the exam. The council also transitioned from a paper-based version of the exam to a computer-based testing (CBT) version. The FE exam now requires you to sit in front of a monitor, solve problems served up by the CBT system, access an electronic reference document, and perform your scratch calculations on a reusable notepad. You may also use an on-screen calculator with which you will likely be unfamiliar. The experience of taking the FE exam will probably be unlike anything you have ever, or will ever again, experience in your career. Similarly, preparing for the exam will be unlike preparing for any other exam.

The CBT FE exam presented three new challenges to me when I began preparing instructional material for it. (1) The subjects in the testable body of knowledge are oddly limited and do not represent a complete cross section of the traditional engineering fundamentals subjects. (2) The NCEES *FE Reference Handbook* (*NCEES Handbook*) is poorly organized, awkwardly formatted, inconsistent in presentation, and idiomatic in convention. (3) Traditional studying, doing homework while working toward a degree, and working at your own desk as a career engineer are poor preparations for the CBT exam experience.

No existing exam review book overcomes all of these challenges. But I wanted you to have something that does. So, in order to prepare you for the CBT FE exam, this book was designed and written from the ground up. In many ways, this book is as unconventional as the exam.

This book covers all of the knowledge areas listed in the NCEES Electrical and Computer FE exam specifications. With the exceptions listed in "How to Use This Book," for better or worse, this book duplicates the terms, variables, and formatting of the *NCEES Handbook* equations.

NCEES has selected what it believes to be all of the engineering fundamentals important to an early-career, minimally qualified engineer, and has distilled them into its single reference, the *NCEES Handbook*. Personally, I cannot accept the premise that engineers learn and use so little engineering while getting their degrees and during their first few career years. However, regardless of whether you accept the NCEES subset of engineering fundamentals, one thing is certain. In serving as your sole source of formulas, theory, methods, and data during the exam, the *NCEES Handbook* severely limits the types of problems that can be included in the FE exam.

The obsolete paper-based exam required very little knowledge outside of what was presented in the previous editions of the *NCEES Handbook*. That *NCEES Handbook* supported a plug-and-chug examinee performance within a constrained body of knowledge. Based on the current FE exam specifications and the *NCEES Handbook*, the CBT FE exam is even more limited than the old paper-based exam. The number (breadth) of knowledge areas, the coverage (depth) of knowledge areas, the number of problems, and the duration of the exam are all significantly reduced. If you are only concerned about passing and/or "getting it over with" before graduation, these reductions are all in your favor. Your only deterrents will be the cost of the exam and the inconvenience of finding a time and place to take it.

Accepting that "it is what it is," I designed this book to guide you through the exam's body of knowledge.

I have several admissions to make: (1) This book contains nothing magical or illicit. (2) This book, by itself, is only one part of a complete preparation. (3) This book stops well short of being perfect. What do I mean by those admissions?

First, this book does not contain anything magical. It's called a "practice problems" book, and though it will save you time in assembling hundreds of practice problems for your review, it will not learn the material for you. Merely owning it is not enough. You will have to put in the "practice" time to use it.

Similarly, there is nothing clandestine or unethical about this book. It does not contain any actual exam problems. It was written in a vacuum, based entirely on the NCEES Electrical and Computer FE exam specifications. This book is not based on feedback from actual examinees.

Truthfully, I expect that many exam problems will be similar to the problems I have used, because NCEES and I developed content with the same set of constraints. (If anything, NCEES is even more constrained when it comes to fringe, outlier, eccentric, or original topics.)

There are a finite number of ways that problems involving Ohm's law ($V = IR$) and Newton's second law of motion ($F = ma$) can be structured. Any similarity between problems in this book and problems in the exam is easily attributed to the limited number of engineering formulas and concepts, the shallowness of the coverage, and the need to keep the entire solution process (reading, researching, calculating, and responding) to less than three minutes for each problem.

Let me give an example to put some flesh on the bones. As any competent engineer can attest, in order to calculate the pressure drop in a pipe network, you would normally have to (1) determine fluid density and viscosity based on the temperature, (2) convert the mass flow rate to a volumetric flow rate, (3) determine the pipe diameter from the pipe size designation (e.g., pipe schedule), (4) calculate the internal pipe area, (5) calculate the flow velocity, (6) determine the specific roughness from the conduit material, (7) calculate the relative roughness, (8) calculate the Reynolds number, (9) calculate or determine the friction factor graphically, (10) determine the equivalent length of fittings and other minor losses, (11) calculate the head loss, and finally, (12) convert the head loss to pressure drop. Length, flow quantity, and fluid property conversions typically add even more complexity. (SSU viscosity? Diameter in inches? Flow rate in SCFM?) As reasonable and conventional as that solution process is, a problem of such complexity is beyond the upper time limit for an FE exam problem.

To make it possible to be solved in the time allowed, any exam problem you see is likely to be more limited. In fact, most or all of the information you need to answer a problem will be given to you in its problem statement. If only the real world were so kind!

Second, by itself, this book is inadequate. It was never intended to define the entirety of your preparation activity. While it introduces essentially all of the exam knowledge areas and content in the *NCEES Handbook*, an introduction is only an introduction. To be a thorough review, this book needs augmentation.

By design, this book has four significant inadequacies.

1. This book has a limited number of pages, so it cannot contain enough of everything for everyone. The number of practice problems that can fit in it is also limited. The number of problems needed by you, personally, to come up to speed in a particular subject may be inadequate. For example, how many problems will you have to review in order to feel comfortable about divergence, curl, differential equations, and linear algebra? (Answer: Probably more than are in all the books you will ever own!) So, additional exposure is inevitable if you want to be adequately prepared in every subject.

2. This book does not contain the *NCEES Handbook*. This book is limited in helping you become familiar with the idiosyncratic sequencing, formatting, variables, omissions, and presentation of topics in the *NCEES Handbook*. The only way to remedy this is to obtain your own copy of the *NCEES Handbook* (available in printed format from PPI and as a free download from the NCEES website) and use it in conjunction with your review.

3. This book does not contain a practice examination (mock exam, sample exam, etc.). With the advent of the CBT format, any sample exam in printed format is little more than another collection of practice problems. The actual FE exam is taken sitting in front of a computer using an online reference book, so the only way to practice is to sit in front of a computer while you answer problems. Using an online reference is very different from the work environment experienced by most engineers, and it will take some getting used to.

4. This book does not contain explanatory background information, including figures and tables of data. Though all problems have associated step-by-step solutions, these solutions will not teach you the underlying engineering principles you need to solve the problems. Trying to extrapolate engineering principles from the solutions is like reading the ending of a book and then trying to guess at the "whos, whats, wheres, whens, and hows." In other words, reviewing solutions is only going to get you so far if you don't understand a topic. To truly understand how to solve practice problems in topics you're unfamiliar with, you'll need an actual review manual like the one PPI publishes, the *FE Electrical and Computer Review Manual*. In it, you'll find all the "whos and whats" you were previously missing and these problems' "endings" will make much more sense.

Third, and finally, I reluctantly admit that I have never figured out how to write or publish a completely flawless first (or even subsequent) edition. The PPI staff comes pretty close to perfection in the areas of design, editing, typography, and illustrating. Subject matter experts help immensely with calculation checking, and beta testing before you see a book helps smooth out wrinkles. However, I still manage to muck up the content. So, I hope you will "let me have it" when you find my mistakes. PPI has established an easy way for you to report an error, as well as to review changes that resulted from errors that others have submitted. Just go to **ppi2pass.com/errata**. When you submit something, I'll receive it via email. When I answer it, you'll receive a response. We'll both benefit.

Best wishes in your examination experience. Stay in touch!

Michael R. Lindeburg, PE

Acknowledgments

Developing a book specific to the computerized Electrical and Computer FE exam has been a monumental project. It involved the usual (from an author's and publisher's standpoint) activities of updating and repurposing existing content and writing new content. However, the project was made extraordinarily more difficult by two factors: (1) a new publishing system, and (2) the publication schedule.

Special thanks go to calculation checkers Ralph Arcena and Nanzhu Zhang; validity reviewers Daniel Blaydon, PE, Andrew Low, PE, James A. Mirabile, PE, Nanzhu Zhang; and problem developers Gregg Wagener, PE, and John A. Camara, PE.

PPI staff members have had a lot of things to say about this book during its development. In reference to you and other examinees being unaware of what PPI staff did, one of the often-heard statements was, "They will never know."

However, I want you to know, so I'm going to tell you.

Director of publishing services Grace Wong managed the gargantuan operation. Production services manager Cathy Schrott kept the process moving smoothly and swiftly, despite technical difficulties that seemed determined to stall the process at every opportunity. Steve Buehler, director of acquisitions, and Nicole Evans, acquisitions editor, arranged for all the outside subject matter experts who were involved with this book. All the content was reviewed for consistency, PPI style, and accuracy by Jennifer Lindeburg King, editor-in-chief.

Though everyone in Publishing Services has a specialty, this project pulled everyone from his or her comfort zone. The entire staff worked on "building" the chapters of this book from scratch, piecing together existing content with new content. Everyone learned (with amazing speed) how to grapple with the complexities of XML and MathML while wrestling misbehaving computer code into submission. Tom Bergstrom, production associate and video production specialist, updated existing illustrations and created new ones. Senior copy editor Scott Marley copy edited the work, and copy editor Robert Genevro proofread, corrected, and paginated.

Consistent with the past 38 years, I continue to thank my wife, Elizabeth, for accepting and participating in a writer's life that is full to overflowing. Even though our children have been out on their own for a long time, we seem to have even less time than we had before. As a corollary to Aristotle's "Nature abhors a vacuum," I propose: "Work expands to fill the void."

To my granddaughter, Sydney, who had to share her Gaga with his writing, I say, "I only worked when you were in school!"

I also appreciate the grant of permission to reproduce materials from several other publishers. In each case, attribution is provided where the material has been included. Neither PPI nor the publishers of the reproduced material make any representations or warranties as to the accuracy of the material, nor are they liable for any damages resulting from its use.

Thank you, everyone! I'm really proud of what you've accomplished. Your efforts will be pleasing to examinees and effective in preparing them for the Electrical and Computer FE exam.

Michael R. Lindeburg, PE

Codes and References Used to Prepare This Book

This book is based on the NCEES *FE Reference Handbook* (*NCEES Handbook*), ninth edition (June 2016 revision). The other documents, codes, and standards that were used to prepare this book were the most current available at the time.

NCEES does not specifically tie the FE exam to any edition (version) of any code or standard. Rather than make the FE exam subject to the vagaries of such codes and standards as are published by the American Chemical Society (ACS), the American Concrete Institute (ACI), the American Institute of Chemical Engineers (AIChE), the American Institute of Steel Construction (AISC), the American National Standards Institute (ANSI), the American Society of Civil Engineers (ASCE), the American Society of Heating, Refrigerating and Air-Conditioning Engineers (ASHRAE), the American Society of Mechanical Engineers (ASME), ASTM International (ASTM), the International Code Council (ICC), the Institute of Electrical and Electronic Engineers (IEEE), the National Fire Protection Association (NFPA), and so on, NCEES effectively writes its own "code," the *NCEES Handbook*.

Most surely, every standard- or code-dependent concept (e.g., flammability) in the *NCEES Handbook* can be traced back to some section of some edition of a standard or code (e.g., 29CFR). So, it would be logical to conclude that you need to be familiar with everything (the limitations, surrounding sections, and commentary) in the code related to that concept. However, that does not seem to be the case. The *NCEES Handbook* is a code unto itself, and you won't need to study the parent documents. Nor will you need to know anything pertaining to related, adjacent, similar, or parallel code concepts. For example, although square concrete columns are covered in the *NCEES Handbook*, round columns are not.

Therefore, although methods and content in the *NCEES Handbook* can be ultimately traced back to some edition (version) of a relevant code, you do not need to know which. You do not need to know whether that content is current, limited in intended application, or relevant. You only need to use the content.

How to Use This Book

This book is written for one purpose, and one purpose only: to get you ready for the FE exam. Because it is a practice problems book, there are a few, but not many, ways to use it. Here's how this book was designed to be used.

GET THE NCEES *FE REFERENCE HANDBOOK*

Get a copy of the NCEES *FE Reference Handbook* (*NCEES Handbook*). Use it as you solve the problems in this book. The *NCEES Handbook* is the only reference you can use during the exam, so you will want to know the sequence of its sections, what data are included, and the approximate locations of important figures and tables in the *NCEES Handbook*. You should also know the terminology (words and phrases) used in the *NCEES Handbook* to describe equations or subjects, because those are the terms you will have to look up during the exam.

The *NCEES Handbook* is available both in printed and PDF format. The index of the print version may help you locate an equation or other information you are looking for, but few terms are indexed thoroughly. The PDF version includes search functionality that is similar to what you'll have available when taking the computer-based exam. In order to find something using the PDF search function, your search term will have to match the content exactly (including punctuation).

There are a few important differences between the ways the *NCEES Handbook* and this book present content. These differences are intentional for the purpose of maintaining clarity and following PPI's publication policies.

- *pressure:* The *NCEES Handbook* primarily uses P for pressure, an atypical engineering convention. This book always uses p so as to differentiate it from P, which is reserved for power, momentum, and axial loading in related chapters.

- *velocity:* The *NCEES Handbook* uses v and occasionally Greek nu, ν, for velocity. This book always uses v to differentiate it from Greek upsilon, υ, which represents specific volume in some topics (e.g., thermodynamics), and Greek nu, ν, which represents absolute viscosity and Poisson's ratio.

- *specific volume:* The *NCEES Handbook* uses v for specific volume. This book always uses Greek upsilon, υ, a convention that most engineers will be familiar with.

- *units:* The *NCEES Handbook* and the FE exam generally do not emphasize the difference between pounds-mass and pounds-force. "Pounds" ("lb") can mean either force or mass. This book always distinguishes between pounds-force (lbf) and pounds-mass (lbm).

WORK THROUGH EVERY PROBLEM

NCEES has greatly reduced the number of subjects about which you are expected to be knowledgeable and has made nothing optional. Skipping your weakest subjects is no longer a viable preparation strategy. You should study all examination knowledge areas, not just your specialty areas. That means you solve every problem in this book and skip nothing. Do not limit the number of problems you solve in hopes of finding enough problems in your areas of expertise to pass the exam.

The FE exam primarily uses SI units. Therefore, the need to work problems in both the customary U.S. and SI systems is greatly diminished. You will need to learn the SI system if you are not already familiar with it.

BE THOROUGH

Being thorough means really doing the work. Some people think they can read a problem statement, think about it for 10 seconds, read the solution, and then say, "Yes, that's what I was thinking of, and that's what I would have done." Sadly, these people find out too late that the human brain doesn't learn very efficiently that way. Under pressure, they find they know and remember very little. For real learning, you'll have to spend some time with a stubby pencil.

There are so many places where you can get messed up solving a problem. Maybe it is in the use of your calculator, like pushing log instead of ln, or forgetting to set the angle to radians instead of degrees, and so on. Maybe it is rusty math. What is $\ln(e^x)$ anyway? How do you factor a polynomial? Maybe it's in finding the data needed or the proper unit conversion. Maybe you're not familiar with the SI system of units. These things take time. And, you have to make the mistakes once so that you do not make them again.

If you do decide to get your hands dirty and actually work these problems, you will have to decide how much reliance you place on this book. It is tempting to turn to a solution when you get slowed down by details or stumped by the subject material. It is tempting to want to maximize the number of problems you solve by spending as little time as possible solving them. However, you need to struggle a little bit more than that to really learn the material.

Studying a new subject is analogous to using a machete to cut a path through a dense jungle. By doing the work, you develop pathways that weren't there before. It is a lot different than just looking at the route on a map. You actually get nowhere by looking at a map. But cut the path once, and you are in business until the jungle overgrowth closes in again. So do the problems—all of them. Do not look at the solutions until you have sweated a little.

1 Units

PRACTICE PROBLEMS

1. What SI unit is equal to the combination of base units $kg \cdot m^2/s^2$?

(A) joule

(B) pascal

(C) tesla

(D) watt

2. What is a kip?

(A) 1000 in-lbf (torque)

(B) 1000 lbm (mass)

(C) 1000 lbf (force)

(D) 1000 psi (pressure)

3. What is a metric ton?

(A) 200 kg

(B) 1000 kg

(C) 2000 kg

(D) 2000 N

SOLUTIONS

1. Kinetic energy is calculated in the SI system as $\frac{1}{2}mv^2$, with units of $kg \cdot m^2/s^2$, which are equal to joules (J).

The answer is (A).

2. The abbreviation *kip* is used for *kilopound*, which is 1000 lbf (pounds of force).

The answer is (C).

3. A metric ton, also known as a *tonne*, is 1000 kg.

The answer is (B).

2 Algebra

PRACTICE PROBLEMS

1. The second and sixth terms of a geometric progression are 3/10 and 243/160, respectively. What is the first term of this sequence?

(A) 1/10

(B) 1/5

(C) 3/5

(D) 3/2

2. Using logarithmic identities, what is most nearly the numerical value for the following expression?

$$\log_3 \frac{3}{2} + \log_3 12 - \log_3 2$$

(A) 0.95

(B) 1.33

(C) 2.00

(D) 2.20

3. Which of the following statements is true for a power series with the general term $a_i x^i$?

I. An infinite power series converges for $x < 1$.

II. Power series can be added together or subtracted within their interval of convergence.

III. Power series can be integrated within their interval of convergence.

(A) I only

(B) II only

(C) I and III

(D) II and III

4. What is the sum of $12 + 13j$ and $7 - 9j$?

(A) $19 - 22j$

(B) $19 + 4j$

(C) $25 - 22j$

(D) $25 + 4j$

5. What is the product of the complex numbers $3 + 4j$ and $7 - 2j$?

(A) $10 + 2j$

(B) $13 + 22j$

(C) $13 + 34j$

(D) $29 + 22j$

SOLUTIONS

1. Use the formula for geometric progression to find the common ratio.

$$\begin{aligned} l &= ar^{n-1} \\ \frac{l_6}{l_2} &= \frac{ar^{6-1}}{ar^{2-1}} = r^4 \\ r &= \sqrt[4]{\frac{l_6}{l_2}} \\ &= \sqrt[4]{\frac{\frac{243}{160}}{\frac{3}{10}}} \\ &= 3/2 \end{aligned}$$

The term before 3/10 is

$$a_1 = \frac{\frac{3}{10}}{\frac{3}{2}} = 1/5$$

The answer is (B).

2. Use the logarithmic identities.

$$\begin{aligned} \log xy &= \log x + \log y \\ \log x/y &= \log x - \log y \end{aligned}$$

$$\begin{aligned} \log_3 \frac{3}{2} + \log_3 12 - \log_3 2 &= \log_3 \frac{\left(\frac{3}{2}\right)(12)}{2} \\ &= \log_3 9 \end{aligned}$$

Since $(3)^2 = 9$,

$$\log_3 9 = 2.00$$

The answer is (C).

3. Power series can be added together, subtracted from each other, differentiated, and integrated within their interval of convergence. The interval of convergence is $-1 < x < 1$.

The answer is (D).

4. Add the real parts and the imaginary parts of each complex number.

$$\begin{aligned} (a+jb)+(c+jd) &= (a+c)+j(b+d) \\ (12+13j)+(7-9j) &= (12+7)+j(13+(-9)) \\ &= 19+4j \end{aligned}$$

The answer is (B).

5. Use the algebraic distributive law and the equivalency $j^2 = -1$.

$$\begin{aligned} (a+jb)(c+jd) &= (ac-bd)+j(ad+bc) \\ (3+4j)(7-2j) &= 21-8j^2+28j-6j \\ &= 21+8+28j-6j \\ &= 29+22j \end{aligned}$$

The answer is (D).

3 Vectors

PRACTICE PROBLEMS

1. What is the name for a vector that represents the sum of two vectors?

(A) scalar

(B) resultant

(C) tensor

(D) moment

2. What is most nearly the length of the resultant of the following vectors?

$$\begin{aligned} &3\mathbf{i} + 4\mathbf{j} - 5\mathbf{k} \\ &7\mathbf{i} + 2\mathbf{j} + 3\mathbf{k} \\ &-16\mathbf{i} - 14\mathbf{j} + 2\mathbf{k} \end{aligned}$$

(A) 3

(B) 4

(C) 10

(D) 14

3. Given the origin-based vector $\mathbf{A} = \mathbf{i} + 2\mathbf{j} + \mathbf{k}$, what is most nearly the angle between $\mathbf{A}$ and the x-axis?

(A) 22°

(B) 24°

(C) 66°

(D) 80°

4. Which is a true statement about these two vectors?

$$\begin{aligned} \mathbf{A} &= \mathbf{i} + 2\mathbf{j} + \mathbf{k} \\ \mathbf{B} &= \mathbf{i} + 3\mathbf{j} - 7\mathbf{k} \end{aligned}$$

(A) Both vectors pass through the point $(0, -1, 3)$.

(B) The vectors are parallel.

(C) The vectors are orthogonal.

(D) The angle between the vectors is 17.4°.

5. What is most nearly the acute angle between vectors $\mathbf{A} = (3, 2, 1)$ and $\mathbf{B} = (2, 3, 2)$, both based at the origin?

(A) 25°

(B) 33°

(C) 35°

(D) 59°

6. Force vectors **A**, **B**, and **C** are applied at a single point.

$$\begin{aligned} \mathbf{A} &= \mathbf{i} + 3\mathbf{j} + 4\mathbf{k} \\ \mathbf{B} &= 2\mathbf{i} + 7\mathbf{j} - \mathbf{k} \\ \mathbf{C} &= -\mathbf{i} + 4\mathbf{j} + 2\mathbf{k} \end{aligned}$$

What is most nearly the magnitude of the resultant force vector, **R**?

(A) 13

(B) 14

(C) 15

(D) 16

SOLUTIONS

1. By definition, the sum of two vectors is known as the resultant.

The answer is (B).

2. The resultant is produced by adding the vectors.

$$\begin{array}{r}
3\mathbf{i} + 4\mathbf{j} - 5\mathbf{k} \\
7\mathbf{i} + 2\mathbf{j} + 3\mathbf{k} \\
-16\mathbf{i} - 14\mathbf{j} + 2\mathbf{k} \\
\hline
-6\mathbf{i} - 8\mathbf{j} + 0\mathbf{k}
\end{array}$$

The length of the resultant vector is

$$\begin{aligned}|\mathbf{R}| &= \sqrt{(-6)^2 + (-8)^2 + (0)^2} \\ &= 10\end{aligned}$$

The answer is (C).

3. The magnitude of vector **A** is

$$|\mathbf{A}| = \sqrt{(1)^2 + (2)^2 + (1)^2} = \sqrt{6}$$

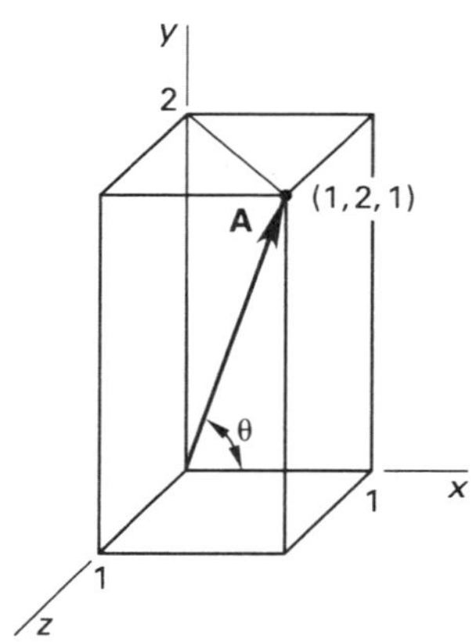

The x-component of the vector is 1, so the direction cosine is

$$\cos\theta_x = \frac{1}{\sqrt{6}}$$

The angle is

$$\theta = \arccos\frac{1}{\sqrt{6}} = 65.9° \quad (66°)$$

The answer is (C).

4. The magnitudes of the two vectors are

$$|\mathbf{A}| = \sqrt{(1)^2 + (2)^2 + (1)^2} = \sqrt{6}$$
$$|\mathbf{B}| = \sqrt{(1)^2 + (3)^2 + (-7)^2} = \sqrt{59}$$

The angle between them is

$$\begin{aligned}\phi &= \arccos\frac{a_x b_x + a_y b_y + a_z b_z}{|\mathbf{A}|\,|\mathbf{B}|} \\ &= \arccos\frac{(1)(1) + (2)(3) + (1)(-7)}{\sqrt{6}\,\sqrt{59}} \\ &= 90°\end{aligned}$$

The vectors are orthogonal.

The answer is (C).

5. The angle between the two vectors is

$$\begin{aligned}\theta &= \arccos\frac{\mathbf{A}\cdot\mathbf{B}}{|\mathbf{A}|\,|\mathbf{B}|} \\ &= \arccos\frac{a_x b_x + a_y b_y + a_z b_z}{|\mathbf{A}|\,|\mathbf{B}|} \\ &= \arccos\frac{(3)(2) + (2)(3) + (1)(2)}{\sqrt{(3)^2 + (2)^2 + (1)^2}\,\sqrt{(2)^2 + (3)^2 + (2)^2}} \\ &= 24.8° \quad (25°)\end{aligned}$$

The answer is (A).

6. The magnitude of **R** is

$$\begin{aligned}|\mathbf{R}| &= \sqrt{(1 + 2 - 1)^2 + (3 + 7 + 4)^2 + (4 - 1 + 2)^2} \\ &= \sqrt{4 + 196 + 25} \\ &= \sqrt{225} \\ &= 15\end{aligned}$$

The answer is (C).

4 Analytic Geometry

PRACTICE PROBLEMS

1. What is the length of the line segment with slope 4/3 that extends from the point (6, 4) to the y-axis?

(A) 10

(B) 25

(C) 50

(D) 75

2. Which of the following equations describes a circle with center at (2, 3) and passing through the point (−3, −4)?

(A) $(x+3)^2+(y+4)^2=85$

(B) $(x+3)^2+(y+2)^2=\sqrt{74}$

(C) $(x-3)^2+(y-2)^2=74$

(D) $(x-2)^2+(y-3)^2=74$

3. The equation for a circle is $x^2+4x+y^2+8y=0$. What are the coordinates of the circle's center?

(A) (−4, −8)

(B) (−4, −2)

(C) (−2, −4)

(D) (2, −4)

4. Which of the following statements is FALSE for all noncircular ellipses?

(A) The eccentricity, e, is less than one.

(B) The ellipse has two foci.

(C) The sum of the two distances from the two foci to any point on the ellipse is $2a$ (i.e., twice the semimajor distance).

(D) The coefficients A and C preceding the x^2 and y^2 terms in the general form of the equation are equal.

5. What is the area of the shaded portion of the circle shown?

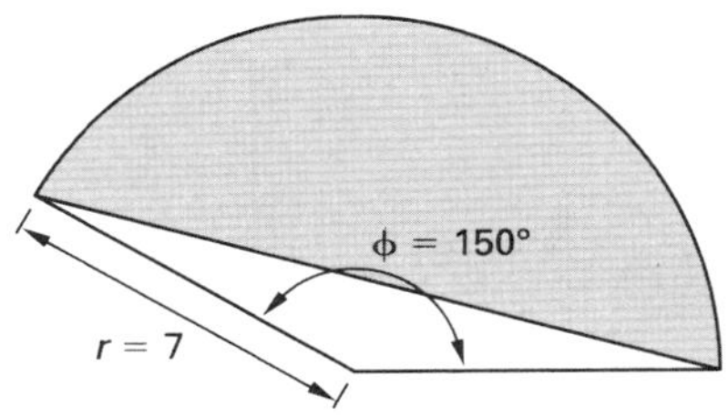

(A) $\frac{5\pi}{6}-1$

(B) $\left(\frac{49}{12}\right)(5\pi-3)$

(C) $\frac{50\pi}{3}$

(D) $49\pi-\sqrt{3}$

6. A pipe with a 20 cm inner diameter is filled to a depth equal to one-third of its diameter. What is the approximate area in flow?

(A) 33 cm²

(B) 60 cm²

(C) 92 cm²

(D) 100 cm²

7. The equation $y=a_1+a_2x$ is an algebraic expression for which of the following?

(A) a cosine expansion series

(B) projectile motion

(C) a circle in polar form

(D) a straight line

8. A circular sector has a radius of 8 cm and an arc length of 13 cm. Most nearly, what is its area?

(A) 48 cm^2

(B) 50 cm^2

(C) 52 cm^2

(D) 60 cm^2

9. The equation $-3x^2 - 4y^2 = 1$ defines

(A) a circle

(B) an ellipse

(C) a hyperbola

(D) a parabola

10. What is the approximate surface area (including both side and base) of a 4 m high right circular cone with a base 3 m in diameter?

(A) 24 m^2

(B) 27 m^2

(C) 32 m^2

(D) 36 m^2

11. What is the approximate area of a circular sector with a radius of 4 and a central angle of 10°?

(A) 0.2

(B) 0.8

(C) 1.4

(D) 2.8

SOLUTIONS

1. The equation of the line is of the form

$$y = mx + b$$

The slope is $m = 4/3$, and a known point is $(x, y) = (6, 4)$. Find the y-intercept, b.

$$4 = \left(\frac{4}{3}\right)(6) + b$$

$$b = 4 - \left(\frac{4}{3}\right)(6) = -4$$

The complete equation is

$$y = \frac{4}{3}x - 4$$

b is the y-intercept, so the intersection with the y-axis is at point $(0, -4)$. The distance between these two points is

$$\begin{aligned} d &= \sqrt{(y_2 - y_1)^2 + (x_2 - x_1)^2} \\ &= \sqrt{(4 - (-4))^2 + (6 - 0)^2} \\ &= 10 \end{aligned}$$

The answer is (A).

2. Substitute the known points into the center-radius form of the equation of a circle.

$$\begin{aligned} r^2 &= (x - h)^2 + (y - k)^2 \\ &= (-3 - 2)^2 + (-4 - 3)^2 \\ &= 74 \end{aligned}$$

The equation of the circle is

$$(x - 2)^2 + (y - 3)^2 = 74$$

$r^2 = 74$, so the radius is $\sqrt{74}$.

The answer is (D).

3. To find the circle's center, put the equation of the circle into standard form.

$$\begin{aligned} x^2 + 4x + y^2 + 8y &= 0 \\ x^2 + 4x + 4 + y^2 + 8y + 16 &= 4 + 16 \\ (x + 2)^2 + (y + 4)^2 &= 20 \end{aligned}$$

The center is at $(-2, -4)$.

The answer is (C).

4. The general form of the equation for an ellipse is

$$Ax^2 + Bxy + Cy^2 + Dx + Ey + F = 0$$

The coefficients preceding the squared terms in the general equation are equal only for a straight line or circle, not for a noncircular ellipse.

The answer is (D).

5. The angle ϕ expressed in radians is

$$\phi = (150°)\left(\frac{2\pi \text{ rad}}{360°}\right) = \frac{5\pi}{6} \text{ rad}$$

The area of the circular segment (the shaded region) is

$$\begin{aligned} A &= \frac{r^2(\phi - \sin\phi)}{2} \\ &= \frac{(7)^2\left(\frac{5\pi}{6} - \sin\frac{5\pi}{6}\right)}{2} \\ &= \left(\frac{49}{2}\right)\left(\frac{5\pi}{6} - \frac{1}{2}\right) \\ &= \left(\frac{49}{12}\right)(5\pi - 3) \end{aligned}$$

The answer is (B).

6. Find the angle ϕ.

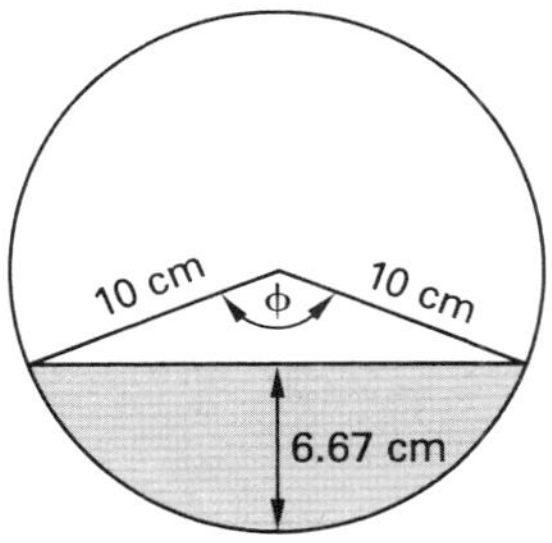

$$\begin{aligned} \phi &= 2\{\arccos[(r - d)/r]\} \\ &= 2\arccos\frac{10 \text{ cm} - 6.67 \text{ cm}}{10 \text{ cm}} \\ &= 2.46 \text{ rad} \end{aligned}$$

Find the area of flow.

$$\begin{aligned} A &= [r^2(\phi - \sin\phi)]/2 \\ &= \frac{(10 \text{ cm})^2(2.46 - \sin 2.46)}{2} \\ &= 91.5 \text{ cm}^2 \quad (92 \text{ cm}^2) \end{aligned}$$

The answer is (C).

7. $y = mx + b$ is the slope-intercept form of the equation of a straight line. a_1 and a_2 are both constants, so $y = a_1 + a_2x$ describes a straight line.

The answer is (D).

8. Find the area of the circular sector.

$$A = sr/2 = \frac{(13 \text{ cm})(8 \text{ cm})}{2} = 52 \text{ cm}^2$$

The answer is (C).

9. The general form of the conic section equation is

$$Ax^2 + Bxy + Cy^2 + Dx + Ey + F = 0$$

$A = -3$, $C = -4$, $F = -1$, and $B = D = E = 0$. A and C are different, so the equation does not define a circle. Calculate the discriminant.

$$B^2 - 4AC = (0)^2 - (4)(-3)(-4) = -48$$

This is less than zero, so the equation defines an ellipse.

The answer is (B).

10. Find the total surface area of a right circular cone. The radius is $r = d/2 = 3 \text{ m}/2 = 1.5 \text{ m}$.

$$\begin{aligned} A &= \text{side area} + \text{base area} = \pi r\left(r + \sqrt{r^2 + h^2}\right) \\ &= \pi(1.5 \text{ m})\left(1.5 \text{ m} + \sqrt{(1.5 \text{ m})^2 + (4 \text{ m})^2}\right) \\ &= 27.2 \text{ m}^2 \quad (27 \text{ m}^2) \end{aligned}$$

The answer is (B).

11. Convert the central angle to radians.

$$(10°)\left(\frac{2\pi}{360°}\right) = 0.175 \text{ rad}$$

Use the formula for the area of a circular sector.

$$A = \phi r^2/2 = \frac{(0.175 \text{ rad})(4)^2}{2} = 1.4$$

The answer is (C).

5 Trigonometry

PRACTICE PROBLEMS

1. To find the width of a river, a surveyor sets up a transit at point C on one river bank and sights directly across to point B on the other bank. The surveyor then walks along the bank for a distance of 275 m to point A. The angle CAB is 57° 28′.

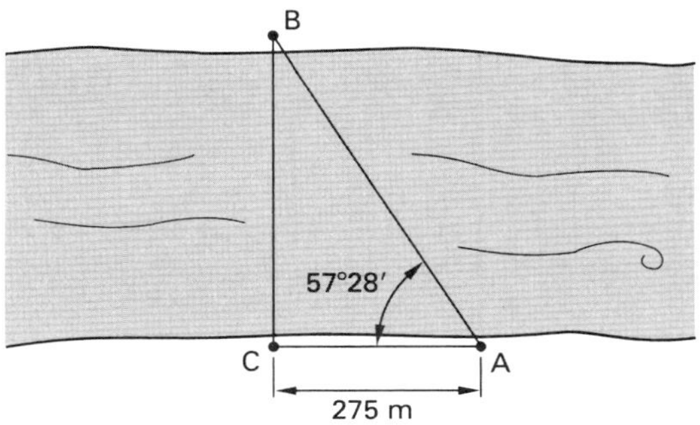

What is the approximate width of the river?

(A) 150 m

(B) 230 m

(C) 330 m

(D) 430 m

2. In the triangle shown, angles ABD and DBC are 90°, AD = 15, DC = 20, and AC = 25.

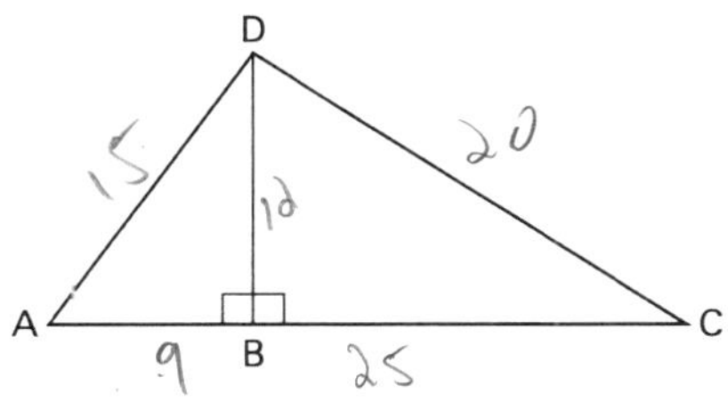

What are the lengths BC and BD, respectively?

(A) 12 and 16

(B) 13 and 17

(C) 16 and 12

(D) 18 and 13

3. Which of the following expressions is equivalent to $\sin 2\theta$?

(A) $2 \sin \theta \cos \theta$

(B) $\cos^2 \theta - \sin^2 \theta$

(C) $\sin \theta \cos \theta$

(D) $\dfrac{1 - \cos 2\theta}{2}$

4. For the right triangle shown, $x = 18$ cm and $y = 13$ cm.

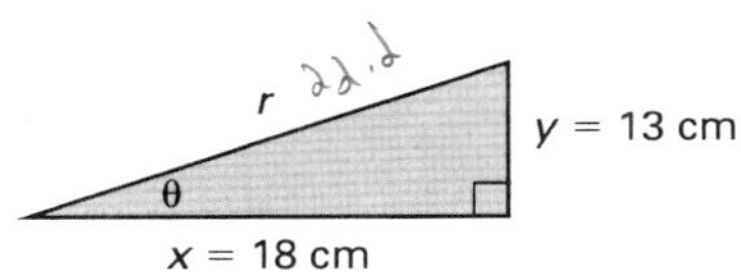

Most nearly, what is $\csc \theta$?

(A) 0.98

(B) 1.2

(C) 1.7

(D) 15

5. A particle moves in the x-y plane. After t seconds, the x- and y-coordinates of the particle's location are $x = 8 \sin t$ and $y = 6 \cos t$. Which of the following equations describes the path of the particle?

(A) $36x^2 + 64y^2 = 2304$

(B) $36x^2 - 64y^2 = 2304$

(C) $64x^2 + 36y^2 = 2304$

(D) $64x^2 - 36y^2 = 2304$

SOLUTIONS

1. Use the formula for the tangent of an angle in a right triangle.

$$\begin{aligned}\tan\theta &= \mathrm{BC}/\mathrm{AC}\\ \mathrm{BC} &= \mathrm{AC}\tan\theta = (275\ \mathrm{m})\tan 57^\circ 28'\\ &= 431.1\ \mathrm{m}\quad (430\ \mathrm{m})\end{aligned}$$

The answer is (D).

2. For right triangle ABD,

$$\begin{aligned}(\mathrm{BD})^2 + (\mathrm{AB})^2 &= (15)^2\\ (\mathrm{BD})^2 &= (15)^2 - (\mathrm{AB})^2\end{aligned}$$

For right triangle DBC,

$$\begin{aligned}(\mathrm{BD})^2 + (25 - \mathrm{AB})^2 &= (20)^2\\ (\mathrm{BD})^2 &= (20)^2 - (25 - \mathrm{AB})^2\end{aligned}$$

Equate the two expressions for $(\mathrm{BD})^2$.

$$\begin{aligned}(15)^2 - (\mathrm{AB})^2 &= (20)^2 - (25)^2 + 50(\mathrm{AB}) - (\mathrm{AB})^2\\ \mathrm{AB} &= \frac{(15)^2 - (20)^2 + (25)^2}{50} = 9\\ \mathrm{BC} &= 25 - \mathrm{AB} = 25 - 9 = 16\\ (\mathrm{BD})^2 &= (15)^2 - (9)^2\\ \mathrm{BD} &= 12\end{aligned}$$

Alternatively, this problem can be solved using the law of cosines.

The answer is (C).

3. The double angle identity is

$$\sin 2\theta = 2\sin\theta\cos\theta$$

The answer is (A).

4. Find the length of the hypotenuse, r.

$$r = \sqrt{x^2 + y^2} = \sqrt{(18\ \mathrm{cm})^2 + (13\ \mathrm{cm})^2} = 22.2\ \mathrm{cm}$$

Find $\csc\theta$.

$$\csc\theta = r/y = \frac{22.2\ \mathrm{cm}}{13\ \mathrm{cm}} = 1.7$$

The answer is (C).

5. Rearrange the two coordinate equations.

$$\begin{aligned}\sin t &= \frac{x}{8}\\ \cos t &= \frac{y}{6}\end{aligned}$$

Use the following trigonometric identity.

$$\begin{aligned}\sin^2\theta + \cos^2\theta &= 1\\ \left(\frac{x}{8}\right)^2 + \left(\frac{y}{6}\right)^2 &= 1\end{aligned}$$

To clear the fractions, multiply both sides by $(8)^2 \times (6)^2 = 2304$.

$$36x^2 + 64y^2 = 2304$$

The answer is (A).

6 Linear Algebra

PRACTICE PROBLEMS

1. What is the solution to the following system of simultaneous linear equations?

$$\begin{aligned} 10x + 3y + 10z &= 5 \\ 8x - 2y + 9z &= 3 \\ 8x + y - 10z &= 7 \end{aligned}$$

(A) $x = 0.326$; $y = -0.192$; $z = 0.586$

(B) $x = 0.148$; $y = 1.203$; $z = 0.099$

(C) $x = 0.625$; $y = 0.186$; $z = -0.181$

(D) $x = 0.282$; $y = -1.337$; $z = -0.131$

2. What is the inverse of matrix **A**?

$$\mathbf{A} = \begin{bmatrix} 2 & 3 \\ 1 & 1 \end{bmatrix}$$

(A) $\begin{bmatrix} 2 & 3 \\ 1 & 1 \end{bmatrix}$

(B) $\begin{bmatrix} 3 & 2 \\ 1 & 1 \end{bmatrix}$

(C) $\begin{bmatrix} 1 & -3 \\ -1 & 2 \end{bmatrix}$

(D) $\begin{bmatrix} -1 & 3 \\ 1 & -2 \end{bmatrix}$

3. If the determinant of matrix **A** is -40, what is the determinant of matrix **B**?

$$\mathbf{A} = \begin{bmatrix} 4 & 3 & 2 & 1 \\ 0 & 1 & 2 & -1 \\ 2 & 3 & -1 & 1 \\ 1 & 1 & 1 & 2 \end{bmatrix} \qquad \mathbf{B} = \begin{bmatrix} 2 & 1.5 & 1 & 0.5 \\ 0 & 1 & 2 & -1 \\ 2 & 3 & -1 & 1 \\ 1 & 1 & 1 & 2 \end{bmatrix}$$

(A) -80

(B) -40

(C) -20

(D) 0.5

SOLUTIONS

1. There are several ways of solving this problem. One is to write the equations in matrix form and solve for the variable matrix, **X**.

$$\mathbf{AX} = \mathbf{B}$$

$$\begin{bmatrix} 10 & 3 & 10 \\ 8 & -2 & 9 \\ 8 & 1 & -10 \end{bmatrix} \begin{bmatrix} x \\ y \\ z \end{bmatrix} = \begin{bmatrix} 5 \\ 3 \\ 7 \end{bmatrix}$$

$$\begin{aligned} \mathbf{AA}^{-1}\mathbf{X} &= \mathbf{A}^{-1}\mathbf{B} \\ \mathbf{IX} &= \mathbf{A}^{-1}\mathbf{B} \\ \mathbf{X} &= \mathbf{A}^{-1}\mathbf{B} \end{aligned}$$

$$\mathbf{X} = \begin{bmatrix} \frac{11}{806} & \frac{20}{403} & \frac{47}{806} \\ \frac{76}{403} & \frac{-90}{403} & \frac{-5}{403} \\ \frac{12}{403} & \frac{7}{403} & \frac{-22}{403} \end{bmatrix} \begin{bmatrix} 5 \\ 3 \\ 7 \end{bmatrix}$$

$$= \begin{bmatrix} (5)\left(\frac{11}{806}\right) + (3)\left(\frac{20}{403}\right) + (7)\left(\frac{47}{806}\right) \\ (5)\left(\frac{76}{403}\right) + (3)\left(\frac{-90}{403}\right) + (7)\left(\frac{-5}{403}\right) \\ (5)\left(\frac{12}{403}\right) + (3)\left(\frac{7}{403}\right) + (7)\left(\frac{-22}{403}\right) \end{bmatrix}$$

$$= \begin{bmatrix} 0.625 \\ 0.186 \\ -0.181 \end{bmatrix}$$

However, substituting the four answer options directly into the original equations is probably the fastest way.

The answer is (C).

2. Find the determinant.

$$|\mathbf{A}| = 2 \times 1 - 1 \times 3 = -1$$

The inverse of a 2 × 2 matrix is

$$\mathbf{A}^{-1} = \frac{\operatorname{adj}(\mathbf{A})}{|\mathbf{A}|} = \frac{\begin{bmatrix} b_2 & -a_2 \\ -b_1 & a_1 \end{bmatrix}}{|\mathbf{A}|}$$

$$= \frac{\begin{bmatrix} 1 & -3 \\ -1 & 2 \end{bmatrix}}{-1}$$

$$= \begin{bmatrix} -1 & 3 \\ 1 & -2 \end{bmatrix}$$

The answer is (D).

3. The first row of matrix **B** is half that of **A**, and the other rows are the same in **A** and **B**, so the determinant of **B** is half the determinant of **A**.

The answer is (C).

7 Calculus

PRACTICE PROBLEMS

1. Which of the following is NOT a correct derivative?

(A) $\frac{d}{dx}\cos x = -\sin x$

(B) $\frac{d}{dx}(1 - x)^3 = -3(1 - x)^2$

(C) $\frac{d}{dx}\frac{1}{x} = -\frac{1}{x^2}$

(D) $\frac{d}{dx}\csc x = -\cot x$

2. What is the derivative, dy/dx, of the equation $x^2y - e^{2x} = \sin y$?

(A) $\frac{2e^{2x}}{x^2 - \cos y}$

(B) $\frac{2e^{2x} - 2xy}{x^2 - \cos y}$

(C) $2e^{2x} - 2xy$

(D) $x^2 - \cos y$

3. What is the approximate area bounded by the curves $y = 8 - x^2$ and $y = -2 + x^2$?

(A) 22

(B) 27

(C) 30

(D) 45

4. What are the minimum and maximum values, respectively, of the equation $f(x) = 5x^3 - 2x^2 + 1$ on the interval $[-2, 2]$?

(A) $-47, 33$

(B) $-4, 4$

(C) $0.95, 1$

(D) $0, 0.27$

5. In vector calculus, a gradient is a

I. vector that points in the direction of a general rate of change of a scalar field

II. vector that points in the direction of the maximum rate of change of a scalar field

III. scalar that indicates the magnitude of the rate of change of a vector field in a general direction

IV. scalar that indicates the maximum magnitude of the rate of change of a vector field in any particular direction

(A) I only

(B) II only

(C) I and III

(D) II and IV

6. Which of the illustrations shown represents the vector field, $\mathbf{F}(x, y) = -y\mathbf{i} + x\mathbf{j}$, for nonzero values of x and y?

(A)

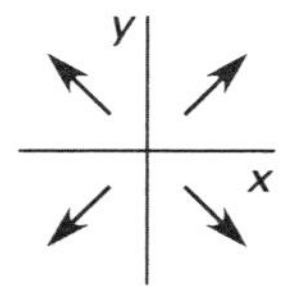

(B)

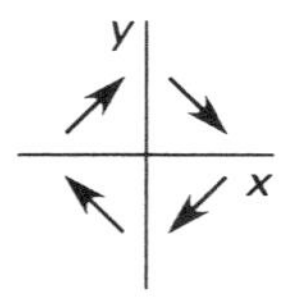

(C)

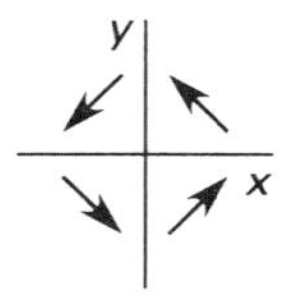

(D)

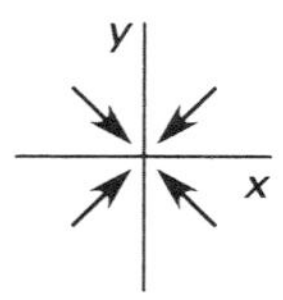

7. If a crop of peaches is picked now, 1000 lugs of peaches will be obtained, which can be sold at \$1.00 per lug. For each week that picking is delayed, the crop will increase by 60 lugs, but the price will drop by \$0.025 per lug. In addition, 10 lugs will spoil for each week of delay. In order to maximize revenue, after how many weeks should the peaches be picked?

(A) 2 weeks

(B) 5 weeks

(C) 7 weeks

(D) 10 weeks

8. Determine the following indefinite integral.

$$\int \frac{x^3 + x + 4}{x^2} dx$$

(A) $\frac{x}{4} + \ln|x| - \frac{4}{x} + C$

(B) $\frac{-x}{2} + \log x - 8x + C$

(C) $\frac{x^2}{2} + \ln|x| - \frac{2}{x^2} + C$

(D) $\frac{x^2}{2} + \ln|x| - \frac{4}{x} + C$

9. Find dy/dx for the parametric equations given.

$$x = 2t^2 - t$$
$$y = t^3 - 2t + 1$$

(A) $3t^2$

(B) $3t^2/2$

(C) $4t - 1$

(D) $(3t^2 - 2)/(4t - 1)$

10. A two-dimensional function, $f(x, y)$, is defined as

$$f(x, y) = 2x^2 - y^2 + 3x - y$$

What is the direction of the line passing through the point $(1, -2)$ that has the maximum slope?

(A) $4\mathbf{i} + 2\mathbf{j}$

(B) $7\mathbf{i} + 3\mathbf{j}$

(C) $7\mathbf{i} + 4\mathbf{j}$

(D) $9\mathbf{i} - 7\mathbf{j}$

11. Evaluate the following limit.

$$\lim_{x \to 2} \frac{x^2 - 4}{x - 2}$$

(A) 0

(B) 2

(C) 4

(D) ∞

12. If $f(x, y) = x^2y^3 + xy^4 + \sin x + \cos^2 x + \sin^3 y$, what is $\partial f/\partial x$?

(A) $(2x + y)y^3 + 3\sin^2 y \cos y$

(B) $(4x - 3y^2)xy^2 + 3\sin^2 y \cos y$

(C) $(3x + 4y^2)xy + 3\sin^2 y \cos y$

(D) $(2x + y)y^3 + (1 - 2\sin x)\cos x$

13. What is dy/dx if $y = (2x)^x$?

(A) $(2x)^x(2 + \ln 2x)$

(B) $2x(1 + \ln 2x)^x$

(C) $(2x)^x(\ln 2x^2)$

(D) $(2x)^x(1 + \ln 2x)$

SOLUTIONS

1. Determine each of the derivatives.

$$\frac{d}{dx}\cos x = -\sin x \quad \text{[OK]}$$

$$\frac{d}{dx}(1-x)^3 = (3)(1-x)^2(-1) = (-3)(1-x)^2 \quad \text{[OK]}$$

$$\frac{d}{dx}\frac{1}{x} = \frac{d}{dx}x^{-1} = (-1)(x^{-2}) = \frac{-1}{x^2} \quad \text{[OK]}$$

$$\frac{d}{dx}\csc x = -\cot x \quad \text{[incorrect]}$$

The answer is (D).

2. Since neither x nor y can be extracted from the equation, rearrange to obtain a homogeneous equation in x and y.

$$x^2y - e^{2x} = \sin y$$
$$f(x, y) = x^2y - e^{2x} - \sin y = 0$$

Take the partial derivatives with respect to x and y.

$$\frac{\partial f(x, y)}{\partial x} = 2xy - 2e^{2x}$$
$$\frac{\partial f(x, y)}{\partial y} = x^2 - \cos y$$

Use implicit differentiation.

$$\frac{\partial y}{\partial x} = \frac{\dfrac{-\partial f(x, y)}{\partial x}}{\dfrac{\partial f(x, y)}{\partial y}} = \frac{2e^{2x} - 2xy}{x^2 - \cos y}$$

The answer is (B).

3. Find the intersection points by setting the two functions equal.

$$-2 + x^2 = 8 - x^2$$
$$2x^2 = 10$$
$$x = \pm\sqrt{5}$$

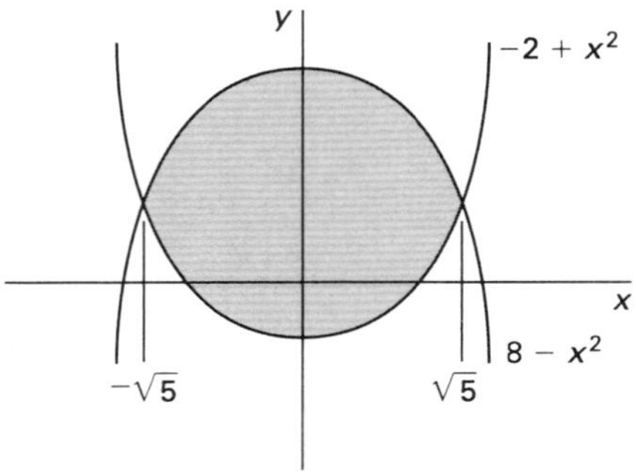

The integral of $f_1(x) - f_2(x)$ represents the area between the two curves between the limits of integration.

$$\begin{aligned} A &= \int_{x_1}^{x_2} (f_1(x) - f_2(x))\,dx \\ &= \int_{-\sqrt{5}}^{\sqrt{5}} ((8 - x^2) - (-2 + x^2))\,dx \\ &= \int_{-\sqrt{5}}^{\sqrt{5}} (10 - 2x^2)\,dx \\ &= \left(10x - \tfrac{2}{3}x^3\right)\Big|_{-\sqrt{5}}^{\sqrt{5}} \\ &= 29.8 \quad (30) \end{aligned}$$

The answer is (C).

4. The critical points are located where the first derivative is zero.

$$\begin{aligned} f(x) &= 5x^3 - 2x^2 + 1 \\ f'(x) &= 15x^2 - 4x \\ 15x^2 - 4x &= 0 \\ x(15x - 4) &= 0 \\ x = 0 \quad \text{or} \quad x &= 4/15 \end{aligned}$$

Test each critical point to determine whether it is a maximum, minimum, or inflection point.

$$\begin{aligned} f''(x) &= 30x - 4 \\ f''(0) &= (30)(0) - 4 \\ &= -4 \end{aligned}$$

This is less than zero, so the critical point at $x = 0$ is a maximum.

$$\begin{aligned} f''\left(\frac{4}{15}\right) &= (30)\left(\frac{4}{15}\right) - 4 \\ &= 4 \end{aligned}$$

This is greater than zero, so the critical point at $x = 4/15$ is a minimum.

These two critical points could be a local maximum and minimum. Compare the values of the function at the critical points with the values of the function at the endpoints.

$$\begin{aligned} f(-2) &= (5)(-2)^3 - (2)(-2)^2 + 1 = -47 \\ f(2) &= (5)(2)^3 - (2)(2)^2 + 1 = 33 \\ f(0) &= (5)(0)^3 - (2)(0)^2 + 1 = 1 \\ f\left(\frac{4}{15}\right) &= (5)\left(\frac{4}{15}\right)^3 - (2)\left(\frac{4}{15}\right)^2 + 1 \\ &= 0.95 \end{aligned}$$

The minimum and maximum values of the equation over the entire interval, −47 and 33, respectively, are at the endpoints.

The answer is (A).

5. A gradient (gradient vector) at some point P is described by use of the gradient (del, grad, nabla, etc.) function, $\nabla f_P \cdot \mathbf{a}$, where $\mathbf{a}$ is a unit vector. In three-dimensional rectangular coordinates, the gradient is equivalent to the partial derivative vector

$$\nabla f \cdot \mathbf{a} = \frac{\partial f}{\partial x}\mathbf{i} + \frac{\partial f}{\partial y}\mathbf{j} + \frac{\partial f}{\partial z}\mathbf{k}$$

This is a vector that points in the direction of the maximum rate of change (i.e., maximum slope).

The answer is (B).

6. From the term $-y\mathbf{i}$, it can be concluded that

(a) for positive values of y, the vector field points to the left

(b) for negative values of y, the vector field points to the right

From the term $+x\mathbf{j}$, it can be concluded that

(a) for positive values of x, the vector field points upward

(b) for negative values of x, the vector field points downward

The answer is (C).

7. Let x represent the number of weeks. The equation describing the price as a function of time is

$$\frac{\text{price}}{\text{lug}} = \$1 - \$0.025x$$

The equation describing the yield is

$$\begin{aligned}\text{lugs sold} &= 1000 + (60 - 10)x \\ &= 1000 + 50x\end{aligned}$$

The revenue function is

$$\begin{aligned}R &= \left(\frac{\text{price}}{\text{lug}}\right)(\text{lugs sold}) \\ &= (1 - 0.025x)(1000 + 50x) \\ &= 1000 + 50x - 25x - 1.25x^2 \\ &= 1000 + 25x - 1.25x^2\end{aligned}$$

To find the maximum of the revenue function, set its derivative equal to zero.

$$\begin{aligned}\frac{dR}{dx} &= 25 - 2.5x = 0 \\ x &= 10 \text{ weeks}\end{aligned}$$

The answer is (D).

8. Separate the fraction into parts and integrate each one.

$$\begin{aligned}\int \frac{x^3 + x + 4}{x^2}dx &= \int \frac{x^3}{x^2}dx + \int \frac{x}{x^2}dx + \int \frac{4}{x^2}dx \\ &= \int x dx + \int \frac{1}{x}dx + 4\int \frac{1}{x^2}dx \\ &= \frac{x^2}{2} + \ln|x| + 4\left(\frac{x^{-1}}{-1}\right) + C \\ &= \frac{x^2}{2} + \ln|x| - \frac{4}{x} + C\end{aligned}$$

The answer is (D).

9. Calculate the derivatives of x and y with respect to t.

$$\begin{aligned}\frac{dy}{dt} &= 3t^2 - 2 \\ \frac{dx}{dt} &= 4t - 1\end{aligned}$$

The derivative of y with respect to x is

$$\begin{aligned}\frac{dy}{dx} &= \frac{\frac{dy}{dt}}{\frac{dx}{dt}} \\ &= \frac{3t^2 - 2}{4t - 1}\end{aligned}$$

The answer is (D).

10. The direction of the line passing through $(1, -2)$ with maximum slope is found by inserting $x = 1$ and $y = -2$ into the gradient vector function.

The gradient of the function is

$$\begin{aligned}\nabla f(x, y, z) &= \frac{\partial f(x, y, z)}{\partial x}\mathbf{i} + \frac{\partial f(x, y, z)}{\partial y}\mathbf{j} + \frac{\partial f(x, y, z)}{\partial z}\mathbf{k} \\ &= \frac{\partial(2x^2 - y^2 + 3x - y)}{\partial x}\mathbf{i} \\ &\quad + \frac{\partial(2x^2 - y^2 + 3x - y)}{\partial y}\mathbf{j} \\ &= (4x + 3)\mathbf{i} - (2y + 1)\mathbf{j}\end{aligned}$$

At $(1, -2)$,

$$\begin{aligned}\nabla f(1, -2) &= ((4)(1) + 3)\mathbf{i} - ((2)(-2) + 1)\mathbf{j} \\ &= 7\mathbf{i} + 3\mathbf{j}\end{aligned}$$

The answer is (B).

11. The expression approaches 0/0 at the limit.

$$\frac{(2)^2 - 4}{2 - 2} = \frac{0}{0}$$

Use L'Hôpital's rule.

$$\lim_{x\to 2}\frac{x^2-4}{x-2} = \lim_{x\to 2}\frac{\frac{d}{dx}(x^2-4)}{\frac{d}{dx}(x-2)} = \lim_{x\to 2}\frac{2x}{1}$$
$$= \frac{(2)(2)}{1}$$
$$= 4$$

This could also be solved by factoring the numerator.

The answer is (C).

12. The partial derivative with respect to x is found by treating all other variables as constants. Therefore, all terms that do not contain x have zero derivatives.

$$\begin{aligned}\frac{\partial f}{\partial x} &= 2xy^3 + y^4 + \cos x + 2\cos x(-\sin x)\\ &= (2x+y)y^3 + (1-2\sin x)\cos x\end{aligned}$$

The answer is (D).

13. From the table of derivatives,

$$\begin{aligned}\mathbf{D}(f(x))^{g(x)} &= g(x)(f(x))^{g(x)-1}\mathbf{D}f(x)\\ &\quad +(\ln f(x))(f(x))^{g(x)}\mathbf{D}g(x)\\ f(x) &= 2x\\ g(x) &= x\\ \frac{d(2x)^x}{dx} &= x(2x)^{x-1}(2) + (\ln 2x)(2x)^x(1)\\ &= (2x)^x + (2x)^x\ln 2x\\ &= (2x)^x(1+\ln 2x)\end{aligned}$$

The answer is (D).

8 Differential Equations

PRACTICE PROBLEMS

1. What is the solution to the following differential equation?

$$y' + 5y = 0$$

(A) $y = 5x + C$

(B) $y = Ce^{-5x}$

(C) $y = Ce^{5x}$

(D) either (A) or (B)

2. What is the solution to the following linear difference equation?

$$(k+1)(y(k+1)) - ky(k) = 1$$

(A) $y(k) = 12 - \dfrac{1}{k}$

(B) $y(k) = 1 - \dfrac{12}{k}$

(C) $y(k) = 12 + 3k$

(D) $y(k) = 3 + \dfrac{1}{k}$

3. What is the general solution to the following differential equation?

$$2\frac{d^2y}{dx^2} - 4\frac{dy}{dx} + 4y = 0$$

(A) $y = C_1 \cos x + C_2 \sin x$

(B) $y = C_1 e^{x} + C_2 e^{-x}$

(C) $y = e^{-x}(C_1 \cos x - C_2 \sin x)$

(D) $y = e^{x}(C_1 \cos x + C_2 \sin x)$

4. What is the general solution to the following differential equation?

$$\frac{d^2y}{dx^2} + 2\frac{dy}{dx} + 2y = 0$$

(A) $y = C_1 \sin x - C_2 \cos x$

(B) $y = C_1 \cos x - C_2 \sin x$

(C) $y = C_1 \cos x + C_2 \sin x$

(D) $y = e^{-x}(C_1 \cos x + C_2 \sin x)$

5. What is the complementary solution to the following differential equation?

$$y'' - 4y' + \tfrac{25}{4}y = 10 \cos 8x$$

(A) $y = 2C_1 x + C_2 x - C_3 x$

(B) $y = C_1 e^{2x} + C_2 e^{1.5x}$

(C) $y = C_1 e^{2x} \cos 1.5x + C_2 e^{2x} \sin 1.5x$

(D) $y = C_1 e^{x} \tan x + C_2 e^{x} \cot x$

6. What is the general solution to the following differential equation?

$$y'' + y' + y = 0$$

(A) $y = e^{-(1/2)x}\left(C_1 \cos \frac{\sqrt{3}}{2}x + C_2 \sin \frac{\sqrt{3}}{2}x\right)$

(B) $y = e^{-(1/2)x}\left(C_1 \cos \frac{3}{2}x + C_2 \sin \frac{3}{2}x\right)$

(C) $y = e^{-2x}\left(C_1 \cos \frac{\sqrt{3}}{2}x + C_2 \sin \frac{\sqrt{3}}{2}x\right)$

(D) $y = e^{-2x}\left(C_1 \cos \frac{3}{2}x + C_2 \sin \frac{3}{2}x\right)$

7. What is the solution to the following differential equation if $x = 1$ at $t = 0$, and $dx/dt = 0$ at $t = 0$?

$$\tfrac{1}{2}\frac{d^2x}{dt^2} + 4\frac{dx}{dt} + 8x = 5$$

(A) $x = e^{-4t} + 4te^{-4t}$

(B) $x = \tfrac{3}{8}e^{-2t}(\cos 2t + \sin 2t) + \dfrac{5}{8}$

(C) $x = e^{-4t} + 4te^{-4t} + \dfrac{5}{8}$

(D) $x = \tfrac{3}{8}e^{-4t} + \tfrac{3}{2}te^{-4t} + \dfrac{5}{8}$

8. In the following differential equation with the initial condition $x(0) = 12$, what is the value of $x(2)$?

$$\frac{dx}{dt} + 4x = 0$$

(A) 3.4×10^{-3}

(B) 4.0×10^{-3}

(C) 5.1×10^{-3}

(D) 6.2×10^{-3}

9. What are the three general Fourier coefficients for the sawtooth wave shown?

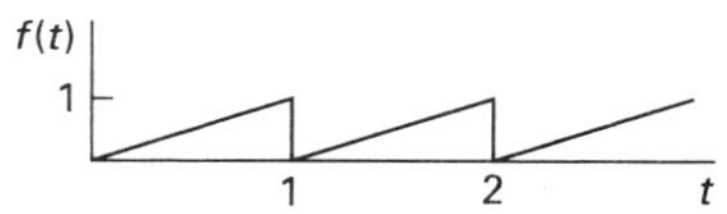

(A) $a_0 = 0,\ a_n = 0,\ b_n = \dfrac{-1}{\pi n}$

(B) $a_0 = \dfrac{1}{2},\ a_n = 0,\ b_n = \dfrac{-1}{\pi n}$

(C) $a_0 = 1,\ a_n = 1,\ b_n = \dfrac{1}{\pi n}$

(D) $a_0 = \dfrac{1}{2},\ a_n = \dfrac{1}{2},\ b_n = \dfrac{1}{\pi n}$

10. The values of an unknown function follow a Fibonacci number sequence. It is known that $f(1) = 4$ and $f(2) = 1.3$. What is $f(4)$?

(A) −4.1

(B) 0.33

(C) 2.7

(D) 6.6

11. A tank with a cross-sectional area of 12 m^2 and a height of 10 m is filled with oil at a rate of 2.2 m^3/min. The density of the oil is 847 kg/m^3. The oil leaks out of the tank from an open tap at the bottom of the tank. The leak rate is $0.11t$ m^3/min, where t is the time in elapsed minutes. Most nearly, what is the level of oil in the tank after the tank has been filled for 6 hr?

(A) 5.2 m

(B) 6.6 m

(C) 7.1 m

(D) 9.4 m

SOLUTIONS

1. This is a first-order linear equation with characteristic equation $r+5=0$. The form of the solution is

$$y = Ce^{-5x}$$

In the preceding equation, the constant, C, could be determined from additional information.

The answer is (B).

2. Since nothing is known about the general form of $y(k)$, the only way to solve this problem is by trial and error, substituting each answer option into the equation in turn. Option B is

$$y(k) = 1 - \frac{12}{k}$$

Substitute this into the difference equation.

$$\begin{aligned}(k+1)(y(k+1)) - k(y(k)) &= 1\\ (k+1)\left(1-\frac{12}{k+1}\right) - k\left(1-\frac{12}{k}\right) &= 1\\ (k+1)\left(\frac{k+1-12}{k+1}\right) - k\left(\frac{k-12}{k}\right) &= 1\\ k+1-12-k+12 &= 1\\ 1 &= 1\end{aligned}$$

$y(k) = 1 - 12/k$ solves the difference equation.

The answer is (B).

3. This is a second-order, homogeneous, linear differential equation. Start by putting it in general form.

$$\begin{aligned}y'' + 2ay' + by &= 0\\ 2y'' - 4y' + 4y &= 0\\ y'' - 2y' + 2y &= 0\\ a &= -2\\ b &= 2\end{aligned}$$

Since $a^2 < 4b$, the form of the equation is

$$\begin{aligned}y &= e^{\alpha x}(C_1 \cos \beta x + C_2 \sin \beta_x)\\ \alpha &= \frac{-a}{2} = \frac{-2}{2} = 1\\ \beta &= \frac{\sqrt{4b-a^2}}{2}\\ &= \frac{\sqrt{(4)(2)-(-2)^2}}{2}\\ &= 1\\ y &= e^{x}(C_1 \cos x + C_2 \sin x)\end{aligned}$$

The answer is (D).

4. The characteristic equation is

$$\begin{aligned}r^2 + 2r + 2 &= 0\\ a &= 2\\ b &= 2\end{aligned}$$

The roots are

$$\begin{aligned}r_{1,2} &= \frac{-a \pm \sqrt{a^2-4b}}{2}\\ &= \frac{-2 \pm \sqrt{(2)^2-(4)(2)}}{2}\\ &= (-1+i),\ (-1-i)\end{aligned}$$

Since $a^2 < 4b$, the solution is

$$\begin{aligned}y &= e^{\alpha x}(C_1 \cos \beta x + C_2 \sin \beta x)\\ \alpha &= \frac{-a}{2} = \frac{-2}{2} = -1\\ \beta &= \frac{\sqrt{4b-a^2}}{2} = \frac{\sqrt{(4)(2)-(2)^2}}{2}\\ &= 1\\ y &= e^{-x}(C_1 \cos x + C_2 \sin x)\end{aligned}$$

The answer is (D).

5. The complementary solution to a nonhomogeneous differential equation is the solution of the homogeneous differential equation.

The characteristic equation is

$$\begin{aligned}r^2 + ar + b &= 0\\ r^2 - 4r + \frac{25}{4} &= 0\end{aligned}$$

So, $a = -4$, and $b = 25/4$.

The roots are

$$\begin{aligned} r_{1,2} &= \frac{-a \pm \sqrt{a^2 - 4b}}{2} \\ &= \frac{-(-4) \pm \sqrt{(-4)^2 - (4)\left(\frac{25}{4}\right)}}{2} \\ &= 2 \pm 1.5i \end{aligned}$$

Since the roots are imaginary, the homogeneous solution has the form of

$$\begin{aligned} y &= e^{\alpha x}(C_1 \cos \beta x + C_2 \sin \beta x) \\ \alpha &= 2 \\ \beta &= \pm 1.5 \end{aligned}$$

The complementary solution is

$$\begin{aligned} y &= e^{2x}(C_1 \cos 1.5x + C_2 \sin 1.5x) \\ &= C_1 e^{2x} \cos 1.5x + C_2 e^{2x} \sin 1.5x \end{aligned}$$

The answer is (C).

6. This is a second-order, homogeneous, linear differential equation with $a = b = 1$. This differential equation can be solved by the method of undetermined coefficients with a solution in the form $y = Ce^{rx}$. The substitution of the solution gives

$$(r^2 + ar + b)Ce^{rx} = 0$$

Because Ce^{rx} can never be zero, the characteristic equation is

$$r^2 + ar + b = 0$$

Because $a^2 = 1 < 4b = 4$, the general solution is in the form

$$y = e^{\alpha x}(C_1 \cos \beta x + C_2 \sin \beta x)$$

Then,

$$\alpha = -a/2 = -1/2$$

$$\beta = \sqrt{\frac{4b - a^2}{2}} = \sqrt{\frac{(4)(1) - (1)^2}{2}} = \frac{\sqrt{3}}{2}$$

Therefore, the general solution is

$$y = e^{-(1/2)x}\left(C_1 \cos \tfrac{\sqrt{3}}{2}x + C_2 \sin \tfrac{\sqrt{3}}{2}x\right)$$

The answer is (A).

7. Multiplying the equation by 2 gives

$$x'' + 8x' + 16x = 10$$

The characteristic equation is

$$r^2 + 8r + 16 = 0$$

The roots of the characteristic equation are

$$r_1 = r_2 = -4$$

The homogeneous (natural) response is

$$x_{\text{natural}} = Ae^{-4t} + Bte^{-4t}$$

By inspection, $x = 5/8$ is a particular solution that solves the nonhomogeneous equation, so the total response is

$$x = Ae^{-4t} + Bte^{-4t} + \frac{5}{8}$$

Since $x = 1$ at $t = 0$,

$$\begin{aligned} 1 &= Ae^0 + \frac{5}{8} \\ A &= \frac{3}{8} \end{aligned}$$

Differentiating x,

$$x' = \frac{3}{8}(-4)e^{-4t} + B(-4te^{-4t} + e^{-4t}) + 0$$

Since $x' = 0$ at $t = 0$,

$$\begin{aligned} 0 &= -\frac{3}{2} + B(0 + 1) \\ B &= \frac{3}{2} \\ x &= \tfrac{3}{8}e^{-4t} + \tfrac{3}{2}te^{-4t} + \frac{5}{8} \end{aligned}$$

The answer is (D).

8. This is a first-order, linear, homogeneous differential equation with characteristic equation $r + 4 = 0$.

$$\begin{aligned} x' - 4x &= 0 \\ x &= x_0 e^{-4t} \\ x(0) &= x_0 e^{(-4)(0)} \\ &= 12 \\ x_0 &= 12 \\ x &= 12e^{-4t} \\ x(2) &= 12e^{(-4)(2)} \\ &= 12e^{-8} \\ &= 4.03 \times 10^{-3} \quad (4.0 \times 10^{-3}) \end{aligned}$$

The answer is (B).

9. By inspection, $f(t) = t$, with the period $T = 1$. The angular frequency is

$$\omega_0 = \frac{2\pi}{T} = \frac{2\pi}{1} = 2\pi$$

The average is

$$\begin{aligned} a_0 &= (1/T)\int_0^T f(t)\,dt = (1/T)\int_0^T t\,dt = \tfrac{1}{2}t^2\Big|_0^1 = \frac{1}{2} - 0 \\ &= \frac{1}{2} \end{aligned}$$

The general a term is

$$\begin{aligned} a_n &= (2/T)\int_0^T f(t)\cos(n\omega_0 t)\,dt \\ &= 2\int_0^1 t\cos(2\pi n t)\,dt \\ &= 0 \end{aligned}$$

The general b term is

$$\begin{aligned} b_n &= (2/T)\int_0^T f(t)\sin(n\omega_0 t)\,dt \\ &= 2\int_0^1 t\sin(2\pi n t)\,dt \\ &= \frac{-1}{\pi n} \end{aligned}$$

The answer is (B).

10. The value of a number in a Fibonacci sequence is the sum of the previous two numbers in the sequence.

Use the second-order difference equation.

$$\begin{aligned} f(k) &= f(k-1) + f(k-2) \\ f(3) &= f(2) + f(1) = 1.3 + 4 \\ &= 5.3 \\ f(4) &= f(3) + f(2) = 5.3 + 1.3 \\ &= 6.6 \end{aligned}$$

The answer is (D).

11. The general equation for the unsteady-state mass balance is

$$\dot{m}_{\text{accumulation}} = \dot{m}_{\text{in}} - \dot{m}_{\text{out}}$$

The mass flow rates can be converted to volumetric flow rates.

$$\begin{aligned} \dot{m}_{\text{accumulation}} &= \dot{m}_{\text{in}} - \dot{m}_{\text{out}} \\ \rho Q_{\text{accumulation}} &= \rho Q_{\text{in}} - \rho Q_{\text{out}} \\ Q_{\text{accumulation}} &= Q_{\text{in}} - Q_{\text{out}} \end{aligned}$$

The volume of oil accumulating in the tank changes with time.

$$\begin{aligned} \frac{dV}{dt} &= Q_{\text{in}} - Q_{\text{out}} \\ &= 2.2\ \frac{\text{m}^3}{\text{min}} - 0.11t\ \frac{\text{m}^3}{\text{min}} \end{aligned}$$

Since the cross-sectional area is constant, the volume of oil accumulating can be expressed in terms of the leak rate.

$$\begin{aligned} A\frac{dh}{dt} &= 2.2\ \frac{\text{m}^3}{\text{min}} - 0.011t\ \frac{\text{m}^3}{\text{min}} \\ \frac{dh}{dt} &= \frac{2.2\ \frac{\text{m}^3}{\text{min}}}{A} - \frac{0.011t\ \frac{\text{m}^3}{\text{min}}}{A} \\ &= \frac{2.2\ \frac{\text{m}^3}{\text{min}}}{12\ \text{m}^2} - \frac{0.011t\ \frac{\text{m}^3}{\text{min}}}{12\ \text{m}^2} \\ &= 0.1833\ \text{m/min} - 0.0009167t\ \text{m/min} \end{aligned}$$

Integrate both sides with respect to time.

$$\int_0^h 1\frac{dh}{dt}\,dt = \int_0^{6\text{ h}}\left(0.1833\ \frac{\text{m}}{\text{min}} - 0.0009167t\ \frac{\text{m}}{\text{min}}\right)dt$$

$$\begin{aligned} h &= \left(0.1833t\ \frac{\text{m}}{\text{min}} - \left(\frac{0.0009167}{2}\right)t^2\ \frac{\text{m}}{\text{min}}\right)\Bigg|_0^{6\text{ h}} \\ &= (0.1833\ \text{m})(6\ \text{h})\left(60\ \frac{\text{min}}{\text{h}}\right) \\ &\quad -\left(\left(\frac{0.0009167}{2}\right)\left((6\ \text{h})\left(60\ \frac{\text{min}}{\text{h}}\right)\right)^2\right) \\ &= 6.585\ \text{m} \quad (6.6\ \text{m}) \end{aligned}$$

The answer is (B).

9 Transforms and Convolution Theory

PRACTICE PROBLEMS

1. The Fourier transform of an impulse $a\delta(t)$ of magnitude a is equal to

(A) a

(B) $a \sin t$

(C) $ae^{j2\pi ft}$

(D) $1/a$

2. The Fourier transform of a discrete-time signal is

(A) always a periodic function

(B) a periodic function only if the signal is periodic

(C) a nonperiodic analog function

(D) a nonperiodic discrete function

3. A discrete linear time-invariant system has an impulse response given by

$$h(n) = \begin{cases} (0.5)^{n-1} & [n \geq 1] \\ 0 & [n < 1] \end{cases}$$

The difference equation representation for the system is

(A) $y(n) = 0.5x(n-1)$

(B) $y(n) = x(n) + 0.5y(n)$

(C) $y(n) = x(n-1) + 0.5y(n-1)$

(D) $y(n) = 0.5x(n) + y(n-1)$

4. A linear time-invariant discrete-time filter has an impulse response given by $h(n) = (-0.5)^n$. What is the filter's steady-state value in response to a unit step?

(A) 0.00

(B) 0.67

(C) 1.0

(D) 2.0

5. A digital filter is observed to have the outputs shown here in response to a unit impulse, $\delta(n)$.

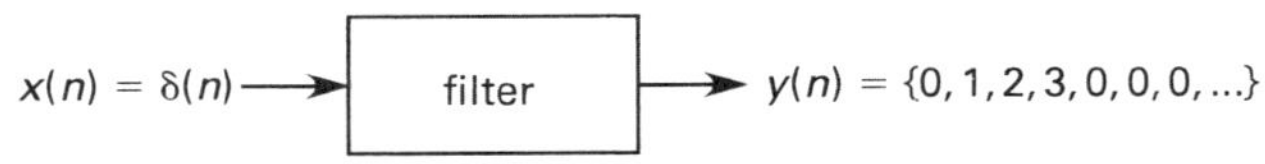

What is the filter's likely response to a unit pulse of three samples (e.g., $x(n) = \{1, 1, 1, 0, 0, 0, ...\}$)?

(A) 0, 1, 2, 3, ...

(B) 0, 1, 3, 6, ...

(C) 1, 1, 1, 0, ...

(D) 1, 2, 3, 0, ...

6. If $x(n)$ is a discrete, imaginary, and even function, its Fourier transform, $X(\omega)$, is

(A) a real and even function

(B) a real and odd function

(C) an imaginary and odd function

(D) an imaginary and even function

7. The transfer function, $H(s)$, and the input function, $X(s)$, for a circuit are

$$H(s) = \frac{2}{s}$$

$$X(s) = \frac{3}{s+5}$$

Using the convolution property, what is the inverse Laplace transform of $H(s)X(s)$ for $t > 0$?

(A) $1.2(1 - e^{-5t})$

(B) $6(e^{5t} - e^{-t})$

(C) $5(e^{5t} - e^{-4t})$

(D) $30(e^{4t} - e^{-t})$

8. Two time-domain functions are

$$x(t) = 2$$

$$y(t) = e^{-t}$$

What is the convolution $x(t) * y(t)$ for $t > 0$?

(A) $1 - e^{-t}$

(B) $2(e^{-t} - 1)$

(C) $2 - e^{-t}$

(D) $2(1 - e^{-t})$

9. The convolution of two continuous time-domain functions is

(A) the Fourier transform of the product of the functions

(B) a function representing the overlap of one function that has been shifted over another

(C) the integral representing the ratio of the area under the functions

(D) a process used to obtain the product of the real roots from functions

10. The function $F(z)$ is a z-transform given by

$$F(z) = \frac{z}{z-5}$$

What is the inverse transform of $F(z)$?

(A) e^{-5i}

(B) $2k$

(C) 2^k

(D) 5^k

11. Two sequences begin as shown.

$$x[n] = 3, 2, 1, 0, \ldots$$
$$y[n] = 1, 2, 3, 0, \ldots$$

The convolution sum of the two sequences is $v[n]$.

$$v[n] = x[n] * y[n] = \sum_{k=-\infty}^{\infty} x[k]y[n-k]$$
$$= \begin{cases} \sum_{k=-\infty}^{\infty} x[k]y[n-k] & [n \geq 0] \\ 0 & [n < 0] \end{cases}$$

What is the value of $v[0]$?

(A) 0

(B) 3

(C) 4

(D) 5

12. A time-domain function is given as

$$f(t) = \sin 3t + 2e^{-3t+4}$$

What is the Laplace transform for this function?

(A) $\frac{s+3}{s^2+9} + \frac{109.2}{s+1}$

(B) $\frac{3}{s^2+9} + \frac{109.2}{s+3}$

(C) $\frac{1}{(s+9)^2} + \frac{55.6}{s+3}$

(D) $\frac{1}{s^2+9} + \frac{2}{s+7}$

13. An s-domain function is

$$F(s) = \frac{4s+8}{(s+2)^2+9}$$

What is the time-domain inverse Laplace transform for this function?

(A) $e^{-t}\cos t$

(B) $2e^{-4t}\cos 3t$

(C) $4e^{-2t}\cos 3t$

(D) $8e^{-2t}\cos 9t$

14. An s-domain function is

$$F(s) = \frac{1}{(s+1)(s+2)}$$

What is the time-domain inverse Laplace transform for this function?

(A) $e^{-t} - e^{-2t}$

(B) $e^{-t} - 2e^{-2t}$

(C) $e^{-t} - te^{-2t}$

(D) $te^{-t} - e^{-2t}$

SOLUTIONS

1. The Fourier transform X of a given signal $x(t)$ is given by

$$X(f) = \int x(t)e^{-j2\pi ft}\,dt$$

Since $x(t) = \delta(t) = 1$ for $t = 0$ s, and $\delta(t) = 0$ elsewhere, then for $x(t) = \delta(t)$, $X(f) = a$.

The answer is (A).

2. The Fourier transform of a discrete-time signal is always a periodic function.

The answer is (A).

3. Let $h'(n) = (0.5)^n$. From a table of z-transform pairs,

$$h'(z) = \frac{1}{1 - 0.5z^{-1}}$$

Using the shift property, since $h(n) = h'(n-1)$, $H(z) = H'(z)z^{-1}$.

$$H(z) = \frac{z^{-1}}{1 - 0.5z^{-1}} = \frac{Y(z)}{X(z)}$$

$$Y(z) - 0.5Y(z)z^{-1} = z^{-1}X(z)$$

Take the inverse z-transform of this expression.

$$y(n) - 0.5y(n-1) = x(n-1)$$

$$y(n) = x(n-1) + 0.5y(n-1)$$

The answer is (C).

4. The filter's z-domain representation as obtained from the transform pair table is

$$H(z) = \frac{1}{1 + 0.5z^{-1}}$$

If the input is a unit step (i.e., $x(n) = u(n)$), then

$$X(z) = \frac{1}{1 - z^{-1}}$$

$$Y(z) = X(z)H(z) = \frac{1}{(1 - z^{-1})(1 + 0.5z^{-1})}$$

Using the final value theorem,

$$\begin{aligned}\lim_{n\to\infty} y(n) &= \lim_{z\to 1}(1 - z^{-1})Y(z) \\ &= (1 - z^{-1})\left(\frac{1}{(1 - z^{-1})(1 + 0.5z^{-1})}\right) \\ &= \lim_{z\to 1}\frac{1}{1 + 0.5z^{-1}} \\ &= \frac{1}{1 + 0.5} \\ &= 0.667 \quad (0.67)\end{aligned}$$

The answer is (B).

5. Obtain $y(n)$ by convolution.

$$\begin{aligned}y(n) &= x(n) * f(n) \\ &= \sum_{k=-\infty}^{\infty} x(k)f(n-k)\end{aligned}$$

$$\begin{aligned}y(0) &= \sum_{k=-\infty}^{\infty} x(k)f(-k) \\ &= x(0)f(0) \\ &= (1)(0) \\ &= 0\end{aligned}$$

$$\begin{aligned}y(1) &= \sum_{k=-\infty}^{\infty} x(k)f(1-k) \\ &= x(0)f(1) + x(1)f(0) \\ &= (1)(1) + (1)(0) \\ &= 1\end{aligned}$$

$$\begin{aligned}y(2) &= \sum_{k=-\infty}^{\infty} x(k)f(2-k) \\ &= x(0)f(2) + x(1)f(1) + x(2)f(0) \\ &= (1)(2) + (1)(1) + (1)(0) \\ &= 3\end{aligned}$$

$$\begin{aligned}y(3) &= \sum_{k=-\infty}^{\infty} x(k)f(3-k) \\ &= x(0)f(3) + x(1)f(2) + x(2)f(1) \\ &= (1)(3) + (1)(2) + (1)(1) \\ &= 6\end{aligned}$$

The answer is (B).

6. The Fourier transform of a discrete signal $x(n)$ is given by

$$\begin{aligned} X(\omega) &= \sum x(n)e^{-j\omega n} \\ &= \sum x(n)(\cos\omega n + j\sin\omega n) \\ &= \sum x(n)\cos\omega n + j\sum x(n)\sin\omega n \end{aligned}$$

In the second term, $x(n)$ is an even function and sin ωn is an odd function, so the entire term is equal to zero, and

$$\begin{aligned} X(\omega) &= \sum x(n)\cos\omega n + 0 \\ &= \sum x(n)\cos\omega n \end{aligned}$$

The function $x(n)$ is imaginary, so $X(\omega)$ is an imaginary function. The functions $x(n)$ and cos ωn are both even, so $X(\omega)$ is an even function.

The answer is (D).

7. The inverse Laplace transforms of the functions $H(s)$ and $X(s)$ are

$$\mathcal{L}^{-1}(H(s)) = h(t) = \mathcal{L}^{-1}\left(\frac{2}{s}\right) = 2$$

$$\mathcal{L}^{-1}(X(s)) = x(t) = \mathcal{L}^{-1}\left(\frac{3}{s+5}\right) = 3e^{-5t}$$

The inverse Laplace transform of the product of two functions in the Laplace domain is the convolution of the time-domain functions.

$$\begin{aligned} v(t) &= h(t) * x(t) = \mathcal{L}^{-1}(H(s)X(s)) \\ &= \int_0^\infty x(t-\tau)h(\tau)\,d\tau \\ &= \int_0^t 3e^{-5(t-\tau)}2\,d\tau \\ &= 6e^{-5t}\int_0^t e^{5t}\,d\tau \\ &= 6e^{-5t}\left(\frac{e^{5t}}{5}\right)\Bigg|_0^t \\ &= \tfrac{6}{5}e^{-5t}(e^{5t}-1) \\ &= 1.2(1-e^{-5t}) \end{aligned}$$

The answer is (A).

8. The convolution integral is

$$\begin{aligned} v(t) = x(t) * y(t) &= \int_0^\infty x(t-\tau)y(\tau)\,d\tau \\ &= \int_0^t 2e^{-\tau}\,d\tau \\ &= 2(e^{-\tau}-1) \end{aligned}$$

The answer is (B).

9. The convolution integral is

$$v(t) = x(t) * y(t) = \int_0^\infty x(t-\tau)y(\tau)\,d\tau$$

A convolution integral can be visualized as the amount of overlap of one function, $x(t)$, as it is shifted by time τ over another function, $y(t)$.

The answer is (B).

10. From a table of z-transforms and their inverses, the inverse transform is

$$F^{-1}(1/1-\beta z^{-1}) = F^{-1}\left(\frac{z}{z-\beta}\right) = \beta^k$$

With β equal to 5, this is

$$F^{-1}(F(z)) = F^{-1}\left(\frac{z}{z-5}\right) = 5^k$$

The answer is (D).

11. At $n=0$, the values of $x[k]$ and $y[0-k]$ for $k=-3$ through 3 are

$$\begin{aligned} x[k] &= \ldots\ 0,0,0,3,2,1,0,\ \ldots \\ y[0-k] &= \ldots\ 0,3,2,1,0,0,0,\ \ldots \end{aligned}$$

The overlap of the input $x[k]$ and the shifted signal $y[n-k]$ occurs at $k=0$ with the values of $x[k]=3$ and $y[n-k]=1$. Their product is $(3)(1)=3$. At all other values of k, either $x[k]$ or $y[n-k]$ is zero, so their product is also zero. Therefore, at $n=0$ the value of $v[0]$ is 3.

The answer is (B).

12. In a table of Laplace transform pairs, find expressions that are equivalent to the terms in the function.

The first term, sin $3t$, is equivalent to the expression e^{-at}

$\sin \beta t$ with $\alpha = 0$ and $\beta = 3$. Use the Laplace transform pair that includes this expression to find the Laplace transform for the first term.

$$e^{-\alpha t}\sin \beta t \rightleftharpoons \frac{\beta}{(s+\alpha)^2+\beta^2}$$

$$e^{0t}\sin 3t \rightleftharpoons \frac{3}{(s+0)^2+(3)^2}$$

$$\sin 3t \rightleftharpoons \frac{3}{s^2+9}$$

The second term can be divided by $2e^4$ to give the expression e^{-at} with $\alpha = 3$. Use the Laplace transform pair that includes this expression to find the Laplace transform for the second term. The factor $2e^4$ is approximately 109.2.

$$e^{-\alpha t} \rightleftharpoons \frac{1}{s+\alpha}$$

$$e^{-3t} \rightleftharpoons \frac{1}{s+3}$$

$$2e^4(e^{-3t}) \rightleftharpoons 2e^4\left(\frac{1}{s+3}\right)$$

$$2e^{-3t+4} \rightleftharpoons \frac{109.2}{s+3}$$

Add the Laplace transforms of the individual terms to get the Laplace transform of the entire function.

$$F(s) = \frac{3}{s^2+9} + \frac{109.2}{s+3}$$

The answer is (B).

13. The inverse Laplace transform can be derived using the Laplace transform pair for $e^{-at}\cos \beta t$ with $\alpha = 2$ and $\beta = 3$.

$$\frac{s+\alpha}{(s+\alpha)^2+\beta^2} \rightleftharpoons e^{-\alpha t}\cos \beta t$$

$$\frac{s+2}{(s+2)^2+(3)^2} \rightleftharpoons e^{-2t}\cos 3t$$

$$F(s) = \frac{4s+8}{(s+2)^2+9} \rightleftharpoons f(t) = 4e^{-2t}\cos 3t$$

The answer is (C).

14. The function $F(s)$ can be decomposed into partial fractions that have known transform pairs. The unknown numerators of the partial fractions are represented by A and B.

$$\begin{aligned} F(s) &= \frac{1}{(s+1)(s+2)} = \frac{A}{s+1} + \frac{B}{s+2} \\ &= \frac{A(s+2)}{(s+1)(s+2)} + \frac{B(s+1)}{(s+1)(s+2)} \\ &= \frac{As+Bs+2A+B}{(s+1)(s+2)} \end{aligned}$$

Because $As+Bs+2A+B$ must equal 1, the two s-terms must cancel (that is, $A+B$ must equal 0) and $2A+B$ must equal 1. Solving the simultaneous equations $A+B=0$ and $2A+B=1$ gives $A=1$ and $B=-1$, so the function can be expressed as

$$F(s) = \frac{1}{s+1} + \frac{-1}{s+2}$$

The inverse Laplace transform can be found using the Laplace transform pair for $1/(s+\alpha)$, with $\alpha = 1$ for the first term and $\alpha = 2$ for the second term.

$$\frac{1}{s+\alpha} \rightleftharpoons e^{-\alpha t}$$

$$F(s) = \frac{1}{s+1} - \frac{1}{s+2} \rightleftharpoons f(t) = e^{-t} - e^{-2t}$$

The answer is (A).

10 Numbering Systems

PRACTICE PROBLEMS

1. What is the 4-bit binary equivalent of the decimal number 13?

(A) 0111

(B) 1011

(C) 1101

(D) 1110

2. Various base-b numbers of the form N_b are combined in an expression.

$$\frac{(1101)_2 - (A)_{16} + (34)_8 + (17)_{10}}{(22)_3}$$

What is the binary equivalent of the evaluated expression?

(A) $(110)_2$

(B) $(1000)_2$

(C) $(10100)_2$

(D) $(110010)_2$

3. What is $(101010111010)_2$ in decimal form?

(A) 1373

(B) 2738

(C) 2746

(D) 5492

4. Most nearly, what is the decimal difference between $(700)_{16}$ and $(700)_8$?

(A) 0

(B) 60

(C) 1300

(D) 5600

5. Most nearly, what is the decimal equivalent of the number $(3620)_7$?

(A) 200

(B) 1300

(C) 3600

(D) 9400

6. What is $(352)_6$ in decimal form?

(A) 100

(B) 140

(C) 180

(D) 350

SOLUTIONS

1. The decimal (base-10) number 13 is the sum of 8, 4, and 1.

$$13_{10} = (1)(2)^3 + (1)(2)^2 + (0)(2)^1 + (1)(2)^0 = (1101)_2$$

The binary equivalent is 1101.

The answer is (C).

2. Find the decimal (base-10) equivalents of all terms in other bases. In the equation for the decimal equivalent of a base-r number, the second term, $\Sigma a_i r^{-i}$, is used for digits to the right of a decimal point; this term is not applicable to any of the numbers in the given expression.

$$\begin{aligned} D &= \sum_{k=0}^{n} a_k r^k + \sum_{i=1}^{m} a_i r^{-i} \\ (1101)_2 &= (1)(2)^0 + (0)(2)^1 + (1)(2)^2 + (1)(2)^3 = 13 \\ (A)_{16} &= (10)(16)^0 = 10 \\ (34)_8 &= (4)(8)^0 + (3)(8)^1 = 28 \\ (22)_3 &= (2)(3)^0 + (2)(3)^1 = 8 \end{aligned}$$

Replace the terms with their decimal equivalents and evaluate the expression.

$$\begin{aligned} \frac{(1101)_2 - (A)_{16} + (34)_8 + (17)_{10}}{(22)_3} &= \frac{13 - 10 + 28 + 17}{8} \\ &= 6 \end{aligned}$$

Convert the result to binary.

$$(6)_{10} = (1)(2)^2 + (1)(2)^1 + (0)(2)^0 = (110)_2$$

The answer is (A).

3. Use the equation for the decimal equivalent of a binary number. Multiply each digit in the binary number by its positional value and add the products.

$$\begin{aligned} D &= a_k 2^k + a_{k-1} 2^{k-1} + \cdots + a_0 + a_{-1} 2^{-1} + \cdots \\ (101010111010)_2 &= (1)(2)^{11} + (0)(2)^{10} + (1)(2)^9 + (0)(2)^8 \\ &\quad + (1)(2)^7 + (0)(2)^6 + (1)(2)^5 + (1)(2)^4 \\ &\quad + (1)(2)^3 + (0)(2)^2 + (1)(2)^1 + (0)(2)^0 \\ &= 2746 \end{aligned}$$

The decimal equivalent is 2746.

The answer is (C).

4. Convert $(700)_{16}$ and $(700)_8$ to decimal (base-10) form.

$$\begin{aligned} D &= \sum_{k=0}^{n} a_k r^k + \sum_{i=1}^{m} a_i r^{-i} \\ (700)_{16} &= (0)(16)^0 + (0)(16)^1 + (7)(16)^2 \\ &= 1792 \\ (700)_8 &= (0)(8)^0 + (0)(8)^1 + (7)(8)^2 \\ &= 448 \end{aligned}$$

Find the decimal difference between $(700)_{16}$ and $(700)_8$.

$$\begin{aligned} (700)_{16} - (700)_8 &= 1792 - 448 \\ &= 1344 \quad (1300) \end{aligned}$$

The answer is (C).

5. Use the formula for the decimal (base-10) equivalent of a base-r number.

$$\begin{aligned} D &= \sum_{k=0}^{n} a_k r^k + \sum_{i=1}^{m} a_i r^{-i} \\ (3620)_7 &= (0)(7)^0 + (2)(7)^1 + (6)(7)^2 + (3)(7)^3 \\ &= 1337 \quad (1300) \end{aligned}$$

The answer is (B).

6. Use the formula for the decimal (base-10) equivalent of a base-r number.

$$\begin{aligned} D &= \sum_{k=0}^{n} a_k r^k + \sum_{i=1}^{m} a_i r^{-i} \\ (352)_6 &= (2)(6)^0 + (5)(6)^1 + (3)(6)^2 \\ &= 140 \end{aligned}$$

The answer is (B).

11 Boolean Algebra

PRACTICE PROBLEMS

(Note: Similar problems are compiled in Chap. 37.)

1. A Boolean function of three variables is defined by

$$F = \overline{A + (B \cdot C)} \cdot C$$

What is an equivalent expression for this function?

(A) $\overline{A} + B + \overline{C}$

(B) $\overline{A} \cdot \overline{B} \cdot C$

(C) $\overline{A} + (B \cdot C) + \overline{C}$

(D) $(\overline{A} \cdot B \cdot C) + \overline{C}$

2. A Boolean function of three variables is defined by

$$F = \overline{(\overline{A} + \overline{C}) \cdot (A \cdot B)}$$

What is an equivalent expression for this function?

(A) $(A \cdot C) + \overline{A} + \overline{B}$

(B) $(A \cdot C) + (B \cdot \overline{C})$

(C) $B + (A \cdot C)$

(D) $A \cdot B \cdot \overline{C}$

3. A Boolean function of five variables is defined by

$$F = (A \cdot B) + (B + C) + ((\overline{A} \cdot D \cdot E) + B) + (A \cdot \overline{B}) + ((\overline{A} \cdot D \cdot E) + (B \cdot C))$$

What is an equivalent expression for this function?

(A) $A + B + C + (\overline{A} \cdot D \cdot E)$

(B) $A + (B \cdot C) + (\overline{A} \cdot D \cdot E)$

(C) $A + C + (\overline{B} \cdot D \cdot E)$

(D) $B + C + (D \cdot E)$

4. A Boolean function of three variables is defined by

$$F = \overline{A \oplus B} + C + B$$

What is an equivalent expression for this function?

(A) $(A \cdot B) + B + C$

(B) $(\overline{A} \cdot \overline{B}) + B + C$

(C) $A + B + C$

(D) $\overline{A} + C$

5. A Boolean function of two variables is defined by

$$F = A \cdot (\overline{A} + B)$$

What is an equivalent expression for this function?

(A) A

(B) $\overline{A}$

(C) $A \cdot B$

(D) $A \oplus B$

SOLUTIONS

1. Use De Morgan's first and second theorems.

$$\begin{aligned} F &= \overline{A + \overline{(B \cdot C)} \cdot C} \\ &= \overline{A} \cdot \overline{\overline{(B \cdot C)}} \cdot C \\ &= \overline{A} \cdot (\overline{B} + \overline{C}) \cdot C \end{aligned}$$

Use the distributive and associative laws.

$$\begin{aligned} F &= \overline{A} \cdot (\overline{B} + \overline{C}) \cdot C \\ &= \overline{A} \cdot ((\overline{B} \cdot C) + (\overline{C} \cdot C)) \\ &= \overline{A} \cdot ((\overline{B} \cdot C) + 0) \\ &= \overline{A} \cdot \overline{B} \cdot C \end{aligned}$$

The answer is (B).

2. Use De Morgan's first and second theorems.

$$\begin{aligned} F &= \overline{\overline{(\overline{A} + \overline{C})} \cdot (A \cdot B)} \\ &= \overline{\overline{(\overline{A} + \overline{C})}} + \overline{(A \cdot B)} \\ &= (\overline{\overline{A}} \cdot \overline{\overline{C}}) + (\overline{A} + \overline{B}) \\ &= (A \cdot C) + \overline{A} + \overline{B} \end{aligned}$$

The answer is (A).

3. Use the associative and commutative laws to restate the function as the sum of products and bring together related terms. The sum of a Boolean expression and itself is equal to itself (that is, $A + A = A$).

$$\begin{aligned} F &= (A \cdot B) + (B + C) + ((\overline{A} \cdot D \cdot E) + B) \\ &\quad + (A \cdot \overline{B}) + ((\overline{A} \cdot D \cdot E) + (B \cdot C)) \\ &= (A \cdot B) + B + C + (\overline{A} \cdot D \cdot E) + B \\ &\quad + (A \cdot \overline{B}) + (\overline{A} \cdot D \cdot E) + (B \cdot C) \\ &= (A \cdot B) + (A \cdot \overline{B}) + B + (B \cdot C) \\ &\quad + C + (\overline{A} \cdot D \cdot E) \end{aligned}$$

Use the distributive law to simplify the expression. The sum of a Boolean expression and one is equal to one ($A + 1 = 1$); the product of a Boolean expression and one is equal to itself ($A \cdot 1 = A$).

$$\begin{aligned} F &= (A \cdot B) + (A \cdot \overline{B}) + B + (B \cdot C) + C + (\overline{A} \cdot D \cdot E) \\ &= (A \cdot (B + \overline{B})) + B + (B \cdot C) + C + (\overline{A} \cdot D \cdot E) \\ &= (A \cdot 1) + B + (B \cdot C) + C + (\overline{A} \cdot D \cdot E) \\ &= A + B + (B \cdot C) + C + (\overline{A} \cdot D \cdot E) \\ &= A + B + (B \cdot C) + (1 \cdot C) + (\overline{A} \cdot D \cdot E) \\ &= A + B + ((B + 1) \cdot C) + (\overline{A} \cdot D \cdot E) \\ &= A + B + (1 \cdot C) + (\overline{A} \cdot D \cdot E) \\ &= A + B + C + (\overline{A} \cdot D \cdot E) \end{aligned}$$

The answer is (A).

4. The EXCLUSIVE-OR or XOR operator is $\oplus$. The expression $A \oplus B$ is defined as equivalent to

$$A \oplus B = (A \cdot \overline{B}) + (\overline{A} \cdot B)$$

Use De Morgan's first and second theorems and the distributive law to evaluate the negated expression $\overline{A \oplus B}$.

$$\begin{aligned} \overline{A \oplus B} &= \overline{(A \cdot \overline{B}) + (\overline{A} \cdot B)} \\ &= \overline{(A \cdot \overline{B})} \cdot \overline{(\overline{A} \cdot B)} \\ &= (\overline{A} + B) \cdot (A + \overline{B}) \\ &= ((\overline{A} + B) \cdot A) + ((\overline{A} + B) \cdot \overline{B}) \\ &= (\overline{A} \cdot A) + (B \cdot A) + (\overline{A} \cdot \overline{B}) + (B \cdot \overline{B}) \\ &= 0 + (B \cdot A) + (\overline{A} \cdot \overline{B}) + 0 \\ &= (A \cdot B) + (\overline{A} \cdot \overline{B}) \end{aligned}$$

Use the associative and distributive laws to simplify the original function.

$$\begin{aligned} F &= \overline{A \oplus B} + C + B \\ &= (A \cdot B) + (\overline{A} \cdot \overline{B}) + C + B \\ &= (A \cdot B) + B + (\overline{A} \cdot \overline{B}) + C \\ &= (A \cdot B) + (1 \cdot B) + (\overline{A} \cdot \overline{B}) + C \\ &= ((A + 1) \cdot B) + (\overline{A} \cdot \overline{B}) + C \\ &= (1 \cdot B) + (\overline{A} \cdot \overline{B}) + C \\ &= B + (\overline{A} \cdot \overline{B}) + C \\ &= (\overline{A} \cdot \overline{B}) + B + C \end{aligned}$$

The answer is (B).

5. Use the distributive law.

$$\begin{aligned} F &= A \cdot (\overline{A} + B) \\ &= (A \cdot \overline{A}) + (A \cdot B) \\ &= 0 + (A \cdot B) \\ &= A \cdot B \end{aligned}$$

The answer is (C).

12 Probability and Statistics

PRACTICE PROBLEMS

1. What is the approximate probability that no people in a group of seven have the same birthday?

(A) 0.056

(B) 0.43

(C) 0.92

(D) 0.94

2. A study gives the following results for a total sample size of 12.

$$3, 4, 4, 5, 8, 8, 8, 10, 11, 15, 18, 20$$

What is most nearly the mean?

(A) 8.9

(B) 9.5

(C) 11

(D) 12

3. A study gives the following results for a total sample size of 8.

$$2, 3, 5, 8, 8, 10, 10, 12$$

The mean of the sample is 7.25. What is most nearly the standard deviation?

(A) 2.5

(B) 2.9

(C) 3.3

(D) 3.7

4. A study gives the following results for a total sample size of 6.

$$10, 12, 13, 14, 14, 15$$

The mean of the sample is 13. What is most nearly the sample standard deviation?

(A) 0.85

(B) 0.90

(C) 1.6

(D) 1.8

5. A study has a sample size of 5, a standard deviation of 10.4, and a sample standard deviation of 11.6. What is most nearly the variance?

(A) 46

(B) 52

(C) 110

(D) 130

6. A study has a sample size of 9, a standard deviation of 4.0, and a sample standard deviation of 4.2. What is most nearly the sample variance?

(A) 16

(B) 18

(C) 34

(D) 36

7. A bag contains 100 balls numbered 1 to 100. One ball is drawn from the bag. What is the probability that the number on the ball selected will be odd or greater than 80?

(A) 0.1

(B) 0.5

(C) 0.6

(D) 0.7

8. Measurements of the water content of soil from a borrow site are normally distributed with a mean of 14.2% and a standard deviation of 2.3%. What is the probability that a sample taken from the site will have a water content above 16% or below 12%?

(A) 0.13

(B) 0.25

(C) 0.37

(D) 0.42

9. What is the probability that either exactly two heads or exactly three heads will be thrown if six fair coins are tossed at once?

(A) 0.35

(B) 0.55

(C) 0.59

(D) 0.63

10. Which of the following statements about probability is NOT valid?

(A) The probability of an event is always positive and within the range of zero and one.

(B) The probability of an event which cannot occur in the population being examined is zero.

(C) If events A and B are mutually exclusive, then the probability of either event occurring in the same population is zero.

(D) The probability of either of two events, A and B, occurring is $P(A+B) = P(A) + P(B) - P(A, B)$.

11. One fair die is used in a dice game. A player wins \$10 if he rolls either a 1 or a 6. He loses \$5 if he rolls any other number. What is the expected winning for one roll of the die?

(A) \$0.00

(B) \$3.30

(C) \$5.00

(D) \$6.70

12. A simulation model for a transportation system is run for 30 replications, and the mean percentage utilization of the transporter used by the system is recorded for each replication. Those 30 data points are then used to form a confidence interval on mean transporter utilization for the system. At a 95% confidence level, the confidence interval is found to be 37.2% ± 3.4%.

Given this information, which of the following facts can be definitively stated about the system?

(A) At 95% confidence, the sample mean of transporter utilization lies in the range 37.2% ± 3.4%.

(B) At 95% confidence, the population mean of transporter utilization lies in the range 37.2% ± 3.4%.

(C) At 95% confidence, the population mean of transporter utilization lies outside of the range 37.2% ± 3.4%.

(D) At 5% confidence, the population mean of transporter utilization lies inside of the range 37.2% ± 3.4%.

13. What is the approximate probability of exactly two people in a group of seven having a birthday on April 15?

(A) 1.2×10^{-18}

(B) 2.4×10^{-17}

(C) 7.4×10^{-6}

(D) 1.6×10^{-4}

14. What are the arithmetic mean and sample standard deviation of the following numbers?

$$71.3, 74.0, 74.25, 78.54, 80.6$$

(A) 74.3, 2.7

(B) 74.3, 3.8

(C) 75.7, 2.7

(D) 75.7, 3.8

15. Four fair coins are tossed at once. What is the probability of obtaining three heads and one tail?

(A) 1/4 (0.25)

(B) 3/8 (0.375)

(C) 1/2 (0.50)

(D) 3/4 (0.75)

16. A manufactured product consists of two parts, A and B, placed end-to-end. The lengths of the parts are normally distributed, with the means and standard deviations shown.

	mean length (cm)	standard deviation (cm)
part A	2.65	0.12
part B	1.45	0.38

Most nearly, what is the probability that the combined length of the two parts is greater than 4.35 cm?

(A) 0.20

(B) 0.26

(C) 0.55

(D) 0.90

17. When introduced, toothpaste brand A took 60% of the market in the first month, the remaining 40% being shared among various competitors. 75% of the people who buy brand A in any given month repeat in the following month. 45% of the customers who purchase a competing brand in any given month switch to brand A in the following month. No brand-switching occurs during the month; all switching occurs at month-end. Most nearly, what is the expected market share of brand A at the end of the third month?

(A) 43%

(B) 51%

(C) 56%

(D) 64%

18. A student's grade in a class is equal to the mean of the student's scores on four homework assignments and a final exam. The final exam is four times as important as each of the homework assignments. The student achieves a score of 68 on the final exam, and scores of 75, 85, 80, and 73 on the homework assignments. What is the student's final score in the class?

(A) 65

(B) 68

(C) 73

(D) 76

19. A data set contains 10 measurements: 1.20 occurs twice, 1.21 occurs once, 1.22 occurs five times, and 1.30 occurs twice. What is the mode of the data set?

(A) 1.20

(B) 1.21

(C) 1.22

(D) 1.23

20. An engineer builds a simulation model of a small factory, runs an experiment with the model using 15 replicates, and determines, at a 90% confidence level, that the mean time an entity spends in the system is 24–28 min. Based on this information, which of the following is a reasonable assumption about the entity time-in-system?

(A) The population mean time-in-system must be in the interval of 24–28 min.

(B) The largest mean time-in-system for any one of the 15 replicates is 28 min.

(C) If the engineer were to run a 16th replicate, the mean time-in-system for that replicate would be in the interval of 24–28 min.

(D) The probability that the population mean time-in-system is greater than 28 min can be estimated as 5%.

21. An engineer is testing the shear strength of spot welds used on a construction site. The engineer's null hypothesis, at a 5% level of significance, is that the mean shear strength of spot welds is at least 3.1 MPa. The engineer randomly selects 15 welds and measures the shear strength of each weld. The engineer finds the sample mean shear strength is 3.07 MPa with a sample standard deviation of 0.069 MPa. Which of the following statements is true?

(A) The null hypothesis should not be rejected.

(B) The null hypothesis should be rejected.

(C) The alternative hypothesis should be rejected.

(D) The null and alternative hypotheses are equally likely.

22. Least-squares linear regression is used to find a straight-line correlation of the form $y = \hat{a} + \hat{b}x$ between compressive strength, x, and intrinsic permeability, y, of various concrete curing rates. Summary quantities are

$$n = 14$$
$$\sum y_i = 572$$
$$\sum y_i^2 = 23{,}530$$
$$\sum x_i = 43$$
$$\sum x_i^2 = 157.42$$
$$\sum x_i y_i = 1697.80$$

The observed actual value of permeability at $x = 3.7$ is $y = 46.1$. Most nearly, what is the absolute value of the corresponding residual?

(A) 1.9

(B) 2.7

(C) 3.5

(D) 6.7

23. A given thermometer always measures temperature as being 2.5°C higher than the actual temperature. The thermometer is used to measure a water sample and finds the temperature is 25°C. The random error of the measurement is known to be zero. Most nearly, what is the true temperature of the sample?

(A) 23°C

(B) 25°C

(C) 28°C

(D) 30°C

24. A 1 L water sample has a known sulfate concentration of 225 mg/L. The sample is separated into ten aliquots, and each is tested separately for sulfate concentration, producing the test results shown.

aliquot	concentration (mg/L)
1	221
2	229
3	221
4	227
5	226
6	223
7	228
8	223
9	222
10	229

Most nearly, what is the random error associated with the laboratory measurements?

(A) 3.0 mg/L

(B) 5.0 mg/L

(C) 10 mg/L

(D) 25 mg/L

25. A linear model of the form $\hat{y} = (\hat{a} + \hat{b}x_i)$ has parameters $\hat{a} = 1.24$ and $\hat{b} = 15.35$. The observed dependent variable, y, has a value of 156.21 at $x = 10$. Most nearly, what is the model residual at $x = 10$?

(A) −1.5

(B) −1.0

(C) 1.0

(D) 1.5

26. Strength tests of aggregate limestone are carried out on three field samples, producing the results shown.

x (lbf/ft^2)	y (lbf/ft^2)	$\hat{y}$ (lbf/ft^2)
27.77	22.65	21.31
56.35	40.22	42.24
112.74	84.23	83.55

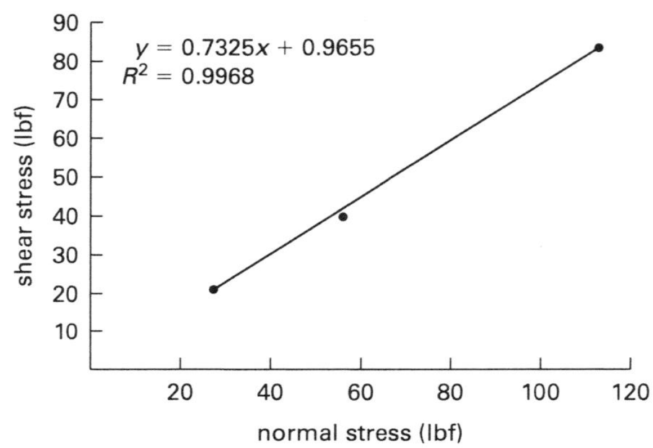

The linear model of the relationship is found to be $\hat{y} = 0.7325x + 0.9655$. Most nearly, what is the mean of the residuals?

(A) −2.02

(B) 0.00

(C) 0.01

(D) 0.68

SOLUTIONS

1. This is the classic "birthday problem." The problem is to find the probability that all seven people have distinctly different birthdays. The solution can be found from simple counting.

The first person considered can be born on any day, which means the probability that the first person will not be born on one of the 365 days of the year is 0.

$$P(1) = 1 - P(\text{not } 1) = 1 - 0 = 1 \quad (365/365)$$

The probability the second person will be born on the same day as the first person is 1 in 365. (The second person can be born on any other of the 364 days.) The probability that the second person is born on any other day is

$$P(2) = 1 - P(\text{not } 2) = 1 - \frac{1}{365} = \frac{364}{365}$$

The third person cannot have been born on either of the same days as the first and second people, which has a 2 in 365 probability of happening. The probability that the third person is born on any other day is

$$P(3) = 1 - P(\text{not } 3) = 1 - \frac{2}{365} = \frac{363}{365}$$

This logic continues to the seventh person. The probability that all seven conditions are simultaneously satisfied is

$$\begin{aligned} &P(7 \text{ distinct birthdays}) \\ &= P(1) \times P(2) \times P(3) \times P(4) \times P(5) \\ &\quad \times P(6) \times P(7) \\ &= \left(\frac{365}{365}\right)\left(\frac{364}{365}\right)\left(\frac{363}{365}\right)\left(\frac{362}{365}\right)\left(\frac{361}{365}\right) \\ &\quad \times \left(\frac{360}{365}\right)\left(\frac{359}{365}\right) \\ &= 0.9438 \quad (0.94) \end{aligned}$$

The answer is (D).

2. The mean is

$$\begin{aligned} \overline{X} &= (1/n)\sum_{i=1}^{n} X_i \\ &= \left(\frac{1}{12}\right)\begin{pmatrix} 3+4+4+5 \\ +8+8+8+10 \\ +11+15+18+20 \end{pmatrix} \\ &= 9.5 \end{aligned}$$

The answer is (B).

3. The standard deviation is calculated using the sample mean as an unbiased estimator of the population mean.

$$\begin{aligned} \sigma &= \sqrt{(1/N)\sum(X_i - \mu)^2} \approx \sqrt{(1/N)\sum(X_i - \overline{X})^2} \\ &= \sqrt{\left(\frac{1}{8}\right)\begin{pmatrix} (2-7.25)^2 + (3-7.25)^2 \\ +(5-7.25)^2 + (8-7.25)^2 \\ +(8-7.25)^2 + (10-7.25)^2 \\ +(10-7.25)^2 + (12-7.25)^2 \end{pmatrix}} \\ &= 3.34 \quad (3.3) \end{aligned}$$

The answer is (C).

4. The sample standard deviation is

$$\begin{aligned} s &= \sqrt{[1/(n-1)]\sum_{i=1}^{n}(X_i - \overline{X})^2} \\ &= \sqrt{\left(\frac{1}{6-1}\right)\begin{pmatrix} (10-13)^2 + (12-13)^2 \\ +(13-13)^2 + (14-13)^2 \\ +(14-13)^2 + (15-13)^2 \end{pmatrix}} \\ &= 1.79 \quad (1.8) \end{aligned}$$

The answer is (D).

5. The variance is the square of the standard deviation.

$$\sigma^2 = (10.4)^2 = 108 \quad (110)$$

The answer is (C).

6. The sample variance is the square of the sample standard deviation.

$$s^2 = (4.2)^2 = 17.64 \quad (18)$$

The answer is (B).

7. There are 50 odd-numbered balls. Including ball 100, there are 20 balls with numbers greater than 80.

$$P(A) = P(\text{ball is odd}) = \frac{50}{100} = 0.5$$

$$P(B) = P(\text{ball} > 80) = \frac{20}{100} = 0.2$$

It is possible for the number on the selected ball to be both odd and greater than 80. Use the law of total probability.

$$\begin{aligned} P(A+B) &= P(A)+P(B)-P(A,B) \\ &= P(A)+P(B)-P(A)P(B) \\ P(\text{odd or} > 80) &= 0.5+0.2-(0.5)(0.2) = 0.6 \end{aligned}$$

The answer is (C).

8. Find the standard normal values for the two points of interest.

$$\begin{aligned} Z_{16\%} &= \frac{x-\mu}{\sigma} = \frac{16\% - 14.2\%}{2.3\%} \\ &= 0.78 \quad [\text{use } 0.80] \end{aligned}$$

$$\begin{aligned} Z_{12\%} &= \frac{x-\mu}{\sigma} = \frac{12\% - 14.2\%}{2.3\%} \\ &= -0.96 \quad [\text{use } -1.00] \end{aligned}$$

Use the unit normal distribution table. The probabilities being sought can be found from the values of $R(x)$ for both standard normal values. $R(0.80) = 0.2119$ and $R(1.00) = 0.1587$. The probability that the sample will fall outside these values is the sum of the two values.

$$\begin{aligned} P(x < 12\% \text{ or } x > 16\%) &= 0.2119 + 0.1587 \\ &= 0.3706 \quad (0.37) \end{aligned}$$

The answer is (C).

9. Find the probability of exactly two heads being thrown. The probability will be the quotient of the total number of possible combinations of six objects taken two at a time and the total number of possible outcomes from tossing six fair coins. The total number of possible outcomes is $(2)^6 = 64$. The total number of possible combinations in which exactly two heads are thrown is

$$\begin{aligned} C(n,r) &= \frac{n!}{r!(n-r)!} = \frac{6!}{2!(6-2)!} \\ &= 15 \end{aligned}$$

The probability of exactly two heads out of six fair coins is

$$P(A) = P(2 \text{ heads}) = \frac{15}{64} = 0.234$$

The probability of exactly three heads being thrown is found similarly. The total number of possible combinations in which exactly three heads are thrown is

$$\begin{aligned} C(n,r) &= \frac{n!}{r!(n-r)!} = \frac{6!}{3!(6-3)!} \\ &= 20 \end{aligned}$$

The probability of exactly three heads out of six fair coins is

$$P(B) = P(3 \text{ heads}) = \frac{20}{64} = 0.313$$

From the law of total probability, the probability that either of these outcomes will occur is the sum of the individual probabilities that the outcomes will occur, minus the probability that both will occur. These two outcomes are mutually exclusive (i.e., both cannot occur), so the probability of both happening is zero.

The total probability is

$$\begin{aligned} P(2 \text{ heads or } 3 \text{ heads}) &= P(A)+P(B)-P(A,B) \\ &= 0.234 + 0.313 - 0 \\ &= 0.547 \quad (0.55) \end{aligned}$$

The answer is (B).

10. If events A and B are mutually exclusive, the probability of both occurring is zero. However, either event could occur by itself, and the probability of that is nonzero.

The answer is (C).

11. For a fair die, the probability of any face turning up is $1/6$. There are two ways to win, and there are four ways to lose. The expected value is

$$\begin{aligned} E[X] &= \sum_{k=1}^{n} x_k f(x_k) = (\$10)\left((2)\left(\frac{1}{6}\right)\right) + (-\$5)\left((4)\left(\frac{1}{6}\right)\right) \\ &= \$0.00 \end{aligned}$$

The answer is (A).

12. A 95% confidence interval on mean transporter utilization means there is a 95% chance the population (or true) mean transporter utilization lies within the given interval.

The answer is (B).

13. Use the binomial probability function to calculate the probability that two of the seven samples will have been born on April 15. $x = 2$, and the sample size, n, is 7.

The probability that a person will have been born on April 15 is 1/365. Therefore, the probability of "success," p, is 1/365, and the probability of "failure," $q = 1 - p$, is 364/365.

$$P_n(x) = \frac{n!}{x!(n-x)!} p^x q^{n-x}$$

$$P_7(2) = \left(\frac{7!}{2!(7-2)!}\right)\left(\frac{1}{365}\right)^2\left(\frac{364}{365}\right)^{7-2}$$

$$= (21)\left(\frac{1}{365}\right)^2\left(\frac{364}{365}\right)^5$$

$$= 1.555 \times 10^{-4} \quad (1.6 \times 10^{-4})$$

The answer is (D).

14. The arithmetic mean is

$$\overline{X} = (1/n)\sum_{i=1}^{n} X_i$$

$$= \left(\frac{1}{5}\right)(71.3 + 74.0 + 74.25 + 78.54 + 80.6)$$

$$= 75.738 \quad (75.7)$$

The sample standard deviation is

$$s = \sqrt{[1/(n-1)]\sum_{i=1}^{n}(X_i - \overline{X})^2}$$

$$= \sqrt{\left(\frac{1}{5-1}\right)\left(\begin{array}{l}(71.3-75.738)^2 + (74.0-75.738)^2 \\ \quad + (74.25-75.738)^2 \\ \quad + (78.54-75.738)^2 \\ \quad + (80.6-75.738)^2\end{array}\right)}$$

$$= 3.756 \quad (3.8)$$

The answer is (D).

15. The binomial probability function can be used to determine the probability of three heads in four trials.

$$p = P(\text{heads}) = 0.5$$
$$q = P(\text{not heads}) = 1 - 0.5 = 0.5$$
$$n = \text{number of trials} = 4$$
$$x = \text{number of successes} = 3$$

From the binomial function,

$$P_n(x) = \frac{n!}{x!(n-x)!} p^x q^{n-x}$$

$$= \left(\frac{4!}{3!(4-3)!}\right)(0.5)^3(0.5)^{4-3}$$

$$= 0.25 \quad (1/4)$$

The answer is (A).

16. Find the mean of the combined length (i.e., add the mean lengths of each part).

$$\mu_L = \mu_1 + \mu_2 = 2.65 \text{ cm} + 1.45 \text{ cm} = 4.1 \text{ cm}$$

The variance of the combined length is the sum of the variances for each part.

$$\sigma_{\text{new}}^2 = \sigma_1^2 + \sigma_2^2 = (0.12 \text{ cm})^2 + (0.38 \text{ cm})^2 = 0.1588 \text{ cm}^2$$

The standard deviation of the combined length is

$$\sigma_L = \sqrt{\sigma_{\text{new}}^2} = \sqrt{0.1588 \text{ cm}^2} = 0.398 \text{ cm}$$

For $P(L > 4.35 \text{ cm})$, the standard normal value is

$$Z = \frac{L - \mu}{\sigma} = \frac{4.35 \text{ cm} - 4.1 \text{ cm}}{0.398 \text{ cm}} = 0.627$$

Interpolating from the unit normal table, the probability $R(0.627)$ of the combined length being greater than 4.35 cm is approximately 0.26.

The answer is (B).

17. Draw a decision tree, and use exhaustive enumeration. The decision tree diagram is

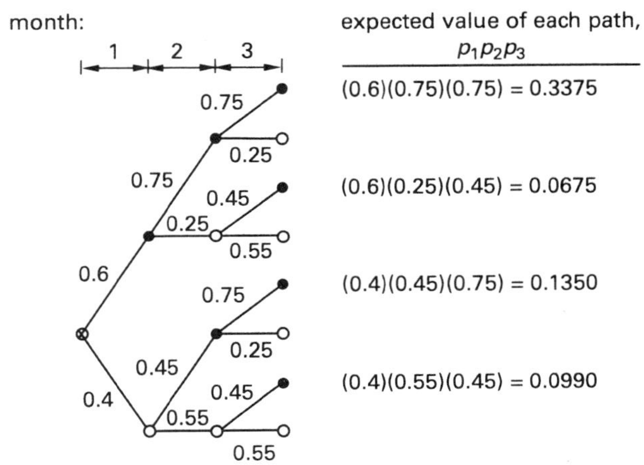

The black nodes represent the expected proportion of people who use brand A at the end of the third month. At that time, the market share of brand A is

$$\begin{aligned} E\{A\} &= \sum p\{A\} = 0.3375 + 0.0675 + 0.1350 + 0.0990 \\ &= 0.639 \quad (64\%) \end{aligned}$$

The answer is (D).

18. The final exam score is more significant than any one of the homework assignment scores. Calculate a weighted arithmetic mean.

$$\begin{aligned} \overline{X}_w &= \frac{\sum w_i X_i}{\sum w_i} = \frac{(75)(1) + (85)(1) + (80)(1) + (73)(1) + (68)(4)}{1 + 1 + 1 + 1 + 4} \\ &= 73 \end{aligned}$$

The answer is (C).

19. The mode is the value that occurs most frequently. 1.22 occurs more times than any other value, so that is the mode.

The answer is (C).

20. A 90% confidence interval for the mean entity time-in-system means there is a 90% chance that the population (or true) mean time-in-system lies within the given interval. Therefore, option A is incorrect. Further, such a confidence interval says nothing either about the individual data values that were used to construct the interval or about future individual experimental values. Therefore, options B and C are incorrect. Because a replicate approach is used in the experiment, the central limit theorem applies, and it can be assumed that the collected data follow a normal distribution. Since the data are normally distributed, it is reasonable to assume that the 10% error (100% – 90%) is equally distributed above and below the confidence interval.

The answer is (D).

21. Since the sample mean and sample standard deviation are used as estimates of the population parameters, use a t-test. The null hypothesis is that

$$\mu \geq 3.1 \text{ MPa}$$

The null hypothesis should be rejected when $t_0 > t_{a,n-1}$ (i.e., when the calculated (absolute) value of t for the null hypothesis is greater than or equal to the value of t as found in the t-distribution table). From the t-distribution table, for a 5% significance and $n-1 = 15 - 1 = 14$ degrees of freedom, the value of t is 1.761.

The actual value of t is

$$\begin{aligned} t &= \frac{\overline{X} - \mu}{\frac{s}{\sqrt{n}}} = \frac{3.07 \text{ MPa} - 3.1 \text{ MPa}}{\frac{0.069 \text{ MPa}}{\sqrt{15}}} \\ &= -1.684 \end{aligned}$$

Since $|-1.684| < |1.76|$, the null hypothesis should not be rejected.

The answer is (A).

22. Least-squares linear regression produces a straight-line correlation equation. The least-squares estimates of the slope and intercept are

$$\begin{aligned} \hat{b} &= \frac{\sum_{i=1}^{n} y_i x_i - \left(\frac{1}{n}\right)\left(\sum_{i=1}^{n} y_i\right)\left(\sum_{i=1}^{n} x_i\right)}{\sum_{i=1}^{n} x_i^2 - \left(\frac{1}{n}\right)\left(\sum_{i=1}^{n} x_i\right)^2} \\ &= \frac{1697.80 - \frac{(572)(43)}{14}}{157.42 - \frac{(43)^2}{14}} \\ &= -2.33 \\ \hat{a} &= \frac{1}{n}\sum_{i=1}^{n} y_i - \frac{\hat{b}}{n}\sum_{i=1}^{n} x_i = \frac{572}{14} - \frac{(-2.33)(43)}{14} \\ &= 48.01 \end{aligned}$$

The equation of the correlation is

$$y = \hat{a} + \hat{b}x = 48.01 - 2.33x$$

The estimated value at $x = 3.7$ is

$$y = \hat{a} + \hat{b}x = 48.01 - (2.33)(3.7) = 39.39$$

The residual value is

$$y_{\text{actual}} - y = 46.1 - 39.39 = 6.71 \quad (6.7)$$

The answer is (D).

23. Use the equation for measurement error, and solve for the true measurement.

$$\begin{aligned} x &= x_{\text{true}} + x_{\text{bias}} + x_{\text{re}} \\ x_{\text{true}} &= x - x_{\text{bias}} - x_{\text{re}} \\ &= 25^\circ\text{C} - 2.5^\circ\text{C} - 0^\circ\text{C} \\ &= 22.5^\circ\text{C} \quad (23^\circ\text{C}) \end{aligned}$$

The answer is (A).

24. Find the mean value of the aliquot testing results.

$$\begin{aligned}\mu &= \frac{\begin{array}{l}221\ \frac{\text{mg}}{\text{L}} + 229\ \frac{\text{mg}}{\text{L}} + 221\ \frac{\text{mg}}{\text{L}} + 227\ \frac{\text{mg}}{\text{L}} \\ +226\ \frac{\text{mg}}{\text{L}} + 223\ \frac{\text{mg}}{\text{L}} + 228\ \frac{\text{mg}}{\text{L}} \\ +223\ \frac{\text{mg}}{\text{L}} + 222\ \frac{\text{mg}}{\text{L}} + 229\ \frac{\text{mg}}{\text{L}}\end{array}}{10} \\ &= 224.9\ \text{mg/L}\end{aligned}$$

The random error is

$$\begin{aligned}\sigma_{\text{population}} = \sigma_{\text{re}} &= \sqrt{\frac{1}{N}\sum(x_i-\mu)^2} \\ &= \sqrt{\frac{1}{10}\left(\begin{array}{l}\left(221\ \frac{\text{mg}}{\text{L}} - 224.9\ \frac{\text{mg}}{\text{L}}\right)^2 \\ +\left(229\ \frac{\text{mg}}{\text{L}} - 224.9\ \frac{\text{mg}}{\text{L}}\right)^2 \\ +\left(221\ \frac{\text{mg}}{\text{L}} - 224.9\ \frac{\text{mg}}{\text{L}}\right)^2 \\ +\left(227\ \frac{\text{mg}}{\text{L}} - 224.9\ \frac{\text{mg}}{\text{L}}\right)^2 \\ +\left(226\ \frac{\text{mg}}{\text{L}} - 224.9\ \frac{\text{mg}}{\text{L}}\right)^2 \\ +\left(223\ \frac{\text{mg}}{\text{L}} - 224.9\ \frac{\text{mg}}{\text{L}}\right)^2 \\ +\left(228\ \frac{\text{mg}}{\text{L}} - 224.9\ \frac{\text{mg}}{\text{L}}\right)^2 \\ +\left(223\ \frac{\text{mg}}{\text{L}} - 224.9\ \frac{\text{mg}}{\text{L}}\right)^2 \\ +\left(221\ \frac{\text{mg}}{\text{L}} - 224.9\ \frac{\text{mg}}{\text{L}}\right)^2 \\ +\left(229\ \frac{\text{mg}}{\text{L}} - 224.9\ \frac{\text{mg}}{\text{L}}\right)^2\end{array}\right)} \\ &= 3.1\ \text{mg/L}\quad (3.0\ \text{mg/L})\end{aligned}$$

The answer is (A).

25. From the equation for the residual,

$$\begin{aligned}e_i &= y_i - \hat{y} = y_i - (\hat{a} + \hat{b}x_i) \\ &= 156.21 - \big(1.24 + (15.35)(10)\big) \\ &= 1.47\quad (1.5)\end{aligned}$$

The answer is (D).

26. From the equation for the residual,

$$\begin{aligned}e_i &= y_i - \hat{y} \\ e_1 &= 22.65\ \frac{\text{lbf}}{\text{ft}^2} - 21.31\ \frac{\text{lbf}}{\text{ft}^2} = 1.34\ \text{lbf/ft}^2 \\ e_2 &= 40.22\ \frac{\text{lbf}}{\text{ft}^2} - 42.24\ \frac{\text{lbf}}{\text{ft}^2} = -2.02\ \text{lbf/ft}^2 \\ e_3 &= 84.23\ \frac{\text{lbf}}{\text{ft}^2} - 83.55\ \frac{\text{lbf}}{\text{ft}^2} = 0.68\ \text{lbf/ft}^2\end{aligned}$$

$$\frac{1}{N}\sum_{i=1}^{N} e_i = \frac{1.34 - 2.02 + 0.68}{3} = 0.00$$

The answer is (B).

Probability/ Statistics

13 Discrete Mathematics

PRACTICE PROBLEMS

1. Set A and set B are subsets of the universal set U. The values within each set are shown.

$$A = \{4, 7, 9\}$$
$$B = \{4, 5, 9, 10\}$$
$$U = \{4, 5, 6, 7, 8, 9, 10\}$$

What is the union of the complement of set A with set B, $\overline{A} \cup B$?

(A) $\{4, 5, 6, 7, 8, 9, 10\}$

(B) $\{4, 5, 7, 9, 10\}$

(C) $\{4, 5, 6, 8, 9, 10\}$

(D) $\{5, 10\}$

2. Set A consists of elements $\{1, 3, 6\}$, and set B consists of elements $\{1, 2, 6, 7\}$. Both sets come from the universal set of $\{1, 2, 3, 4, 5, 6, 7, 8\}$. What is the intersection of the complement of set A with set B, $\overline{A} \cap B$?

(A) $\{2, 7\}$

(B) $\{2, 3, 7\}$

(C) $\{2, 4, 5, 7, 8\}$

(D) $\{4, 5, 8\}$

3. The output, $y[k]$, of the second-order difference equation shown is represented as an alternating stream of ones and zeros. The input function, $x[k]$, is zero for all values of k.

$$y[k] + a_1 y[k-1] + a_2 y[k-2] = x[k]$$

k	0	1	2	3	4	5	6	7	8	9
$y[k]$	1	0	1	0	1	0	1	0	1	0

What are the values of a_1 and a_2?

(A) $a_1 = 0$, $a_2 = -1$

(B) $a_1 = 0$, $a_2 = 0$

(C) $a_1 = 1$, $a_2 = -1$

(D) $a_1 = 1$, $a_2 = 0$

4. A universal set, U, contains the elements a, b, c, d, e, f, g, h, i, j, k, l, m, and n. Sets A and B are subsets of U. Set A contains a, b, c, d, e, and f, and set B contains e, f, g, h, i, and j.

What set is denoted by $(A \cup B) \cap \overline{(A \cap B)}$?

(A) $\{e, f, g, h, i, j\}$

(B) $\{a, b, c, d, k, l, m, n\}$

(C) $\{a, b, c, d, g, h, i, j\}$

(D) $\{e, f, k, l, m, n\}$

5. In each of the four illustrations shown, the shaded portion represents the result of an operation performed on two sets, A and B, within a universal set, U.

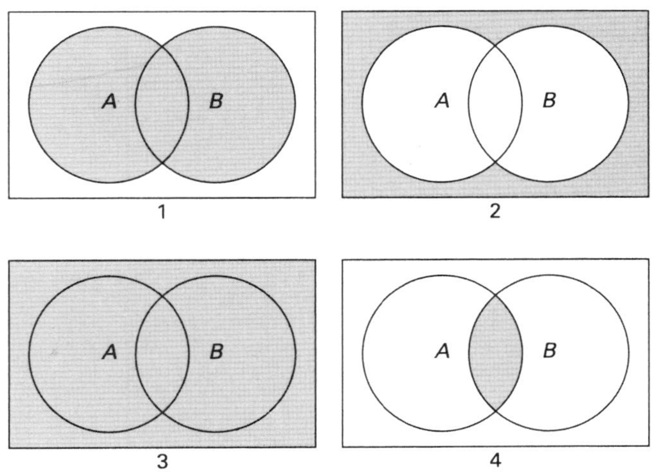

In order from 1 to 4, the four illustrations represent

(A) U, $\overline{A} \cup \overline{B}$, $\varnothing$, and $A \cap B$

(B) $A \cup B$, $\overline{A} \cap \overline{B}$, U, and $A \cap B$

(C) U, $\overline{A} \cap \overline{B}$, $\varnothing$, and $\overline{A} \cup \overline{B}$

(D) $A \cup B$, $\overline{A} \cup \overline{B}$, U, and $A \cup B$

6. Which of these four equations in set theory is FALSE?

(A) $\overline{(A \cup \overline{A})} = \varnothing$

(B) $\overline{A \cap B} = \overline{A} \cap \overline{B}$

(C) $\overline{A} \cup (B \cap C) = (\overline{A} \cup B) \cap (\overline{A} \cup C)$

(D) $(A \cap \overline{B}) \cap C = A \cap (\overline{B} \cap C)$

7. In the Venn diagram shown, the universal set, U, represents the general population. Set A represents people born in leap years, set B represents people with pet snakes, and set C represents people who have red hair.

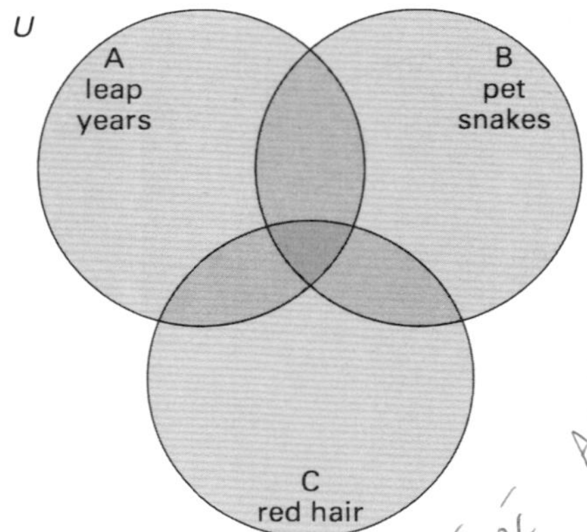

Which group of people within the general population is represented by the expression $A \cap ((\overline{B \cap C}) \cup (\overline{B} \cap A))$?

(A) everyone born in a leap year who has a pet snake or red hair, but not both

(B) everyone born in a leap year except those who have both a pet snake and red hair

(C) everyone who has red hair or a pet snake

(D) everyone who doesn't have red hair or doesn't have a pet snake

8. A set is defined by the expression

$$(A \cap A) \cup (B \cap A) \cup (A \cap C) \cup (B \cap C)$$

What is an equivalent expression for this set?

(A) $A \cup (B \cap C)$

(B) $A \cup (B \cup C)$

(C) $A \cap (B \cup C)$

(D) $A \cap (B \cap C)$

9. In the Venn diagram shown, the universal set, U, represents the general population. Set A represents people born in leap years, set B represents people with pet snakes, and set C represents people who have red hair.

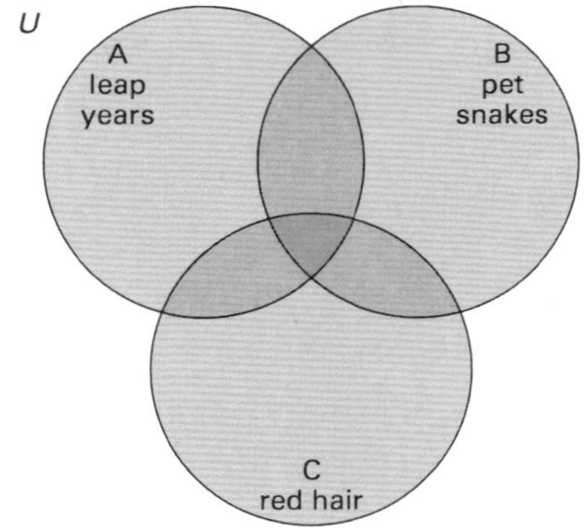

Which of the following expressions represents the set of everyone who either was born in a leap year or has both a pet snake and red hair?

(A) $A \cap B \cap C$

(B) $A \cup B \cup C$

(C) $A \cap (B \cup C)$

(D) $A \cup (B \cap C)$

10. Among all those who took an examination, set A represents people who scored 60% to 70%, set B represents people who scored 65% to 85%, and set C represents people who scored 80% to 90%. Which of the following expressions represents the set of everyone who scored 80% to 85%?

(A) $A \cup (B \cap C)$

(B) $A \cap (B \cup C)$

(C) $(A \cup B) \cap C$

(D) $(A \cap B) \cup C$

11. A universal set, U, contains the elements a, b, c, d, e, f, g, h, i, j, k, l, m, and n. Sets A and B are subsets of U. Set A contains a, b, c, d, e, and f, and set B contains e, f, g, h, i, and j.

What set is denoted by $\overline{A \cup B}$?

(A) $\{a, b, c, d, g, h, i, j\}$

(B) $\{e, f\}$

(C) $\{e, f, k, l, m, n\}$

(D) $\{k, l, m, n\}$

12. In each of the four illustrations shown, the shaded portion represents the result of an operation performed on two sets, A and B, within a universal set, U.

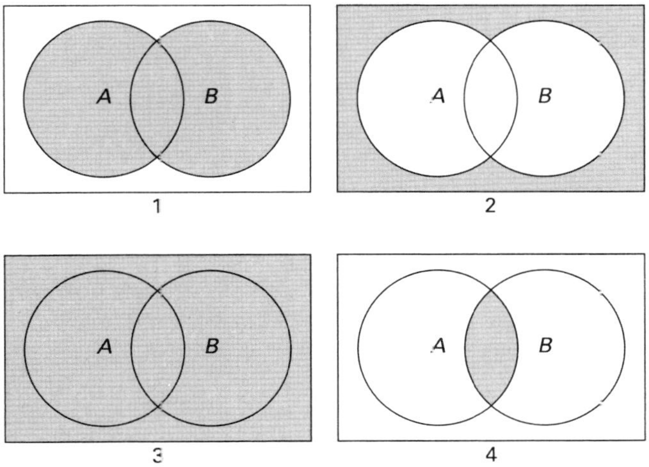

Which statement about these illustrations is FALSE?

(A) Illustration 1 represents $(A \cap \overline{B}) \cup B$.

(B) Illustration 2 represents $(\overline{A} \cup B) \cap \overline{B}$.

(C) Illustration 3 represents $A \cup \overline{(A \cap B)}$.

(D) Illustration 4 represents $B \cap (A \cup B)$.

13. A loan of \$10,000 is made at a 15% effective interest per payment period. Each loan payment totals \$3000. After the first payment is made, what is the outstanding balance on the loan?

(A) \$7000

(B) \$8050

(C) \$8500

(D) \$9000

14. The performance of a recursive filter is described by the first-order difference equation.

$$y[k] - \tfrac{1}{2}y[k-1] = x[k]$$

$x[k]$ is given by $x[k] = C\delta(k)$ where C is an arbitrary complex number. The auxiliary or initial condition is described by $y[-1] = a$.

What is the value of $y[1]$?

(A) $\frac{1}{2}(C + \frac{1}{2}a)$

(B) $C + \frac{1}{2}a$

(C) 1/2

(D) $C + a$

15. As it pertains to polynomials and difference equations, what does the term *second order* indicate?

(A) a relationship among three consecutive values

(B) an equation represented by a polynomial whose largest exponent is 2

(C) both (A) and (B)

(D) neither (A) nor (B)

SOLUTIONS

1. The complement of set A contains all of the members of set U that are not members of set A: $\{5, 6, 8, 10\}$.

The union of the complement of set A with set B is the set of all members appearing in either.

$$\begin{aligned}\overline{A} \cup B &= \{5,\ 6,\ 8,\ 10\} \cup \{4,\ 5,\ 9,\ 10\} \\ &= \{4,\ 5,\ 6,\ 8,\ 9,\ 10\}\end{aligned}$$

The answer is (C).

2. The complement of set A consists of all elements in the universal set that are not in set A: $\{2, 4, 5, 7, 8\}$.

The intersection of the complement of set A with set B is the set of all elements appearing in both.

$$\begin{aligned}\overline{A} \cap B &= \{2,\ 4,\ 5,\ 7,\ 8\} \cap \{1,\ 2,\ 6,\ 7\} \\ &= \{2,\ 7\}\end{aligned}$$

The answer is (A).

3. Since the value of $y[k-2]$ is needed from the table, start with $k = 2$, making $y[k-2] = y[0]$, the first entry in the table.

Insert the values from the table for $k = 2$.

$$\begin{aligned}y[2] + a_1 y[1] + a_2 y[0] &= x[2] \\ 1 + a_1(0) + a_2(1) &= 0 \\ a_2 &= -1\end{aligned}$$

Similarly, insert values from the table for $k = 3$.

$$\begin{aligned}y[3] + a_1 y[2] + a_2 y[1] &= x[3] \\ 0 + a_1(1) + (-1)(0) &= 0 \\ a_1 &= 0\end{aligned}$$

The answer is (A).

4. The union of A and B contains every element that is in either A or B.

$$A \cup B = \{a, b, c, d, e, f, g, h, i, j\}$$

The intersection of A and B contains every element that is in both A and B.

$$A \cap B = \{e, f\}$$

The complement of the intersection of A and B contains every element that is in U and is not in $A \cap B$.

$$\overline{A \cap B} = \{a, b, c, d, g, h, i, j, k, l, m, n\}$$

The intersection of $A \cup B$ and $\overline{A \cap B}$ contains every element that is in both $A \cup B$ and $\overline{A \cap B}$.

$$(A \cup B) \cap \overline{(A \cap B)} = \{a, b, c, d, g, h, i, j\}$$

The answer is (C).

5. The shaded portion in illustration 1 shows the union of A and B, $A \cup B$. This set contains every point that is either in A or in B (or both).

The shaded portion in illustration 2 shows the intersection of the complements of A and B, $\overline{A} \cap \overline{B}$. This set contains every point that is both not in A and not in B.

The shaded portion in illustration 3 shows the universal set, U. This set contains every point.

The shaded portion in illustration 4 shows the intersection of A and B, $A \cap B$. This set contains every point that is both in A and in B.

The answer is (B).

6. Option B is an erroneous version of one of De Morgan's laws. The law correctly stated is $\overline{A \cap B} = \overline{A} \cup \overline{B}$.

Option A follows from the complement laws. The union of A and not-A is the universal set, U; the complement of U is the null set.

Option C follows from the distributive law.

Option D follows from the associative law.

The answer is (B).

7. Analyze the components of the expression to determine which areas of the diagram are included.

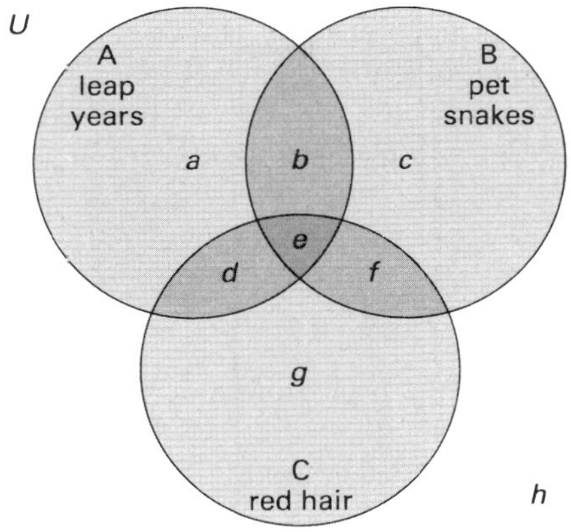

$$B \cap C = \{e, f\}$$
$$\overline{B \cap C} = \{a, b, c, d, g, h\}$$
$$\overline{B} \cap A = \{a, d\}$$
$$(\overline{B \cap C}) \cup (\overline{B} \cap A) = \{a, b, c, d, g, h\}$$
$$A \cap ((\overline{B \cap C}) \cup (\overline{B} \cap A)) = \{a, b, d\}$$

The set represented by the expression includes a, b, and d, so it represents the set of everyone born in a leap year except those who have both a pet snake and red hair.

The answer is (B).

8. The intersection of a set with itself is the set, so

$$A \cap A = A$$

All the members of $B \cap A$ and $A \cap C$ must also be members of the set A, so

$$A \cup (B \cap A) \cup (A \cap C) = A$$

Therefore,

$$\begin{aligned}(A \cap A) &\cup (B \cap A) \cup (A \cap C) \cup (B \cap C) \\ &= A \cup (B \cap A) \cup (A \cap C) \cup (B \cap C) \\ &= A \cup (B \cap C)\end{aligned}$$

The answer is (A).

9. The desired set includes everyone in set A as well as everyone who is in both sets B and C. This is expressed by

$$A \cup (B \cap C)$$

The answer is (D).

10. $A \cup B$ includes all those who scored 60% to 85%, so $(A \cup B) \cap C$ includes only those who scored 80% to 85%. Option C is correct.

$B \cap C$ includes only those who scored 80% to 85%, so $A \cup (B \cap C)$ includes those who scored 60% to 70% as well as those who scored 80% to 85%. Option A is wrong.

$B \cup C$ includes all those who scored 65% to 90%, so $A \cap (B \cup C)$ includes only those who scored 65% to 70%. Option B is wrong.

$A \cap B$ includes only those who scored 65% to 70%, so $(A \cup B) \cap C$ includes those who scored 65% to 70% as well as those who scored 80% to 90%. Option D is wrong.

The answer is (C).

11. The union of A and B is the set of all elements contained in either set.

$$A \cup B = \{a, b, c, d, e, f, g, h, i, j\}$$

The complement of $A \cup B$ is the set of all elements within the universal set U that are not contained in $A \cup B$.

$$\overline{A \cup B} = \{k, l, m, n\}$$

The answer is (D).

12. $A \cup B$ is the set $\{a, b, c\}$, so $B \cap (A \cup B)$ is the set $\{b, c\}$. Option D is false.

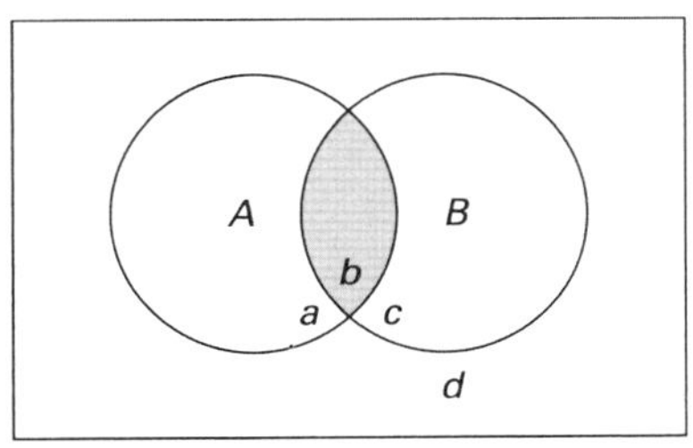

$A \cap \overline{B}$ is the set $\{a\}$, so $(A \cap \overline{B}) \cup B$ is the set $\{a, b, c\}$. Option A is true.

$\overline{A} \cup B$ is the set $\{b, c, d\}$, so $(\overline{A} \cup B) \cap \overline{B}$ is the set $\{d\}$. Option B is true.

$A \cap B$ is the set $\{b\}$, so $\overline{A \cap B}$ is the set $\{a, c, d\}$ and $A \cup \overline{(A \cap B)}$ is the set $\{a, b, c, d\}$. Option C is true.

The answer is (D).

13. Use the first-order difference equation. The balance at the end of period k after making a payment of A is

$$P_k = P_{k-1}(1 + i) - A$$

When $k = 1$, the balance is

$$\begin{aligned} P_1 &= P_0(1+i) - A \\ &= (\$10{,}000)(1+0.15) - \$3000 \\ &= \$8500 \end{aligned}$$

The answer is (C).

14. Rearrange the difference equation.

$$y[k] = x[k] + \tfrac{1}{2}y[k-1]$$

Find $y[0]$.

$$\begin{aligned} y[0] &= x[0] + \tfrac{1}{2}y[0-1] \\ &= C + \tfrac{1}{2}a \end{aligned}$$

Knowing $y[0]$, find $y[1]$.

$$\begin{aligned} y[1] &= x[1] + \tfrac{1}{2}y[1-1] \\ &= x[1] + \tfrac{1}{2}y[0] \\ &= \tfrac{1}{2}\left(C + \tfrac{1}{2}a\right) \end{aligned}$$

The answer is (A).

15. The term *second order* indicates both a polynomial of order 2, and, for a difference equation, a relationship that is among three consecutive values or points.

The answer is (C).

14 Types of Materials

PRACTICE PROBLEMS

1. Which of the following is a thermocouple used to measure?

(A) heat of fusion at a junction

(B) absolute temperature at a junction

(C) temperature at a junction relative to a reference temperature

(D) potential difference between two junctions

2. Refer to the phase diagram shown.

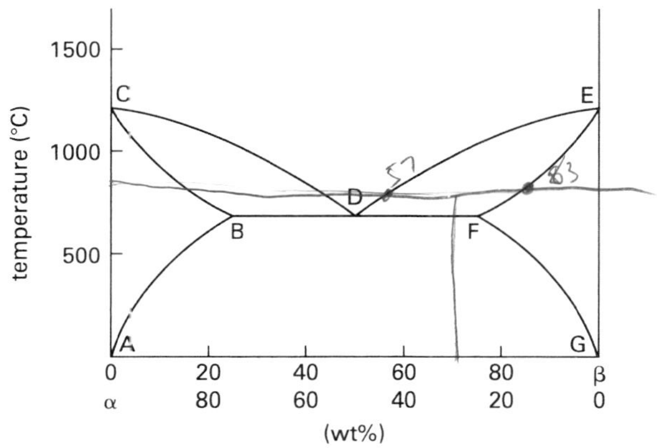

Approximately how much solid (as a percentage by weight) exists when the mixture is 30% α and 70% β and the temperature is 800°C?

(A) 0%

(B) 19%

(C) 30%

(D) 50%

3. Which of the following characteristics describes martensite?

I. high ductility

II. formed by quenching austenite

III. high hardness

(A) I only

(B) I and II

(C) I and III

(D) II and III

4. Which of the following processes can increase the deformation resistance of steel?

I. tempering

II. hot working

III. adding alloying elements

IV. hardening

(A) I and II

(B) I and IV

(C) II and III

(D) III and IV

5. What is the hardest form of steel?

(A) pearlite

(B) ferrite

(C) bainite

(D) martensite

6. A mixture of ice and water is held at a constant temperature of 0°C. How many degrees of freedom does the mixture have?

(A) −1

(B) 0

(C) 1

(D) 2

7. Refer to the phase diagram shown.

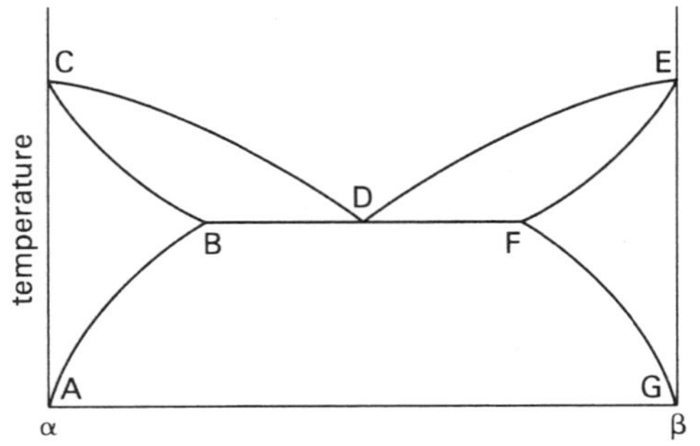

The liquidus line is

(A) CDFG

(B) CDE

(C) CBFE

(D) ABDE

8. Refer to the phase diagram shown.

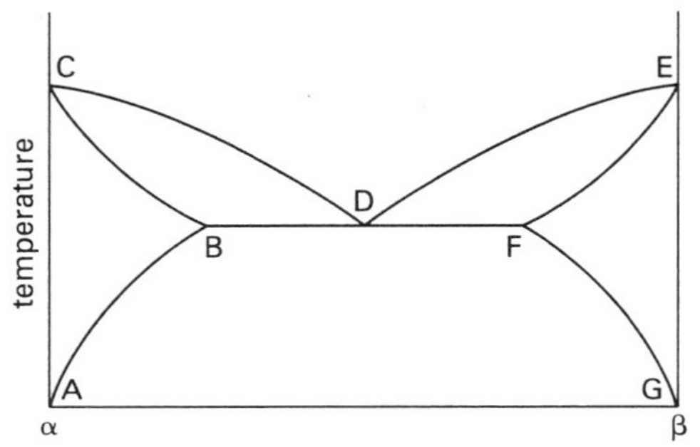

The region enclosed by points DEF can be described as a

(A) mixture of solid β component and liquid α component

(B) mixture of solid β component and liquid β component

(C) peritectic composition

(D) mixture of solid β component and a molten mixture of α and β components

9. Which of the following figures is a cooling curve of a pure metal?

(A)

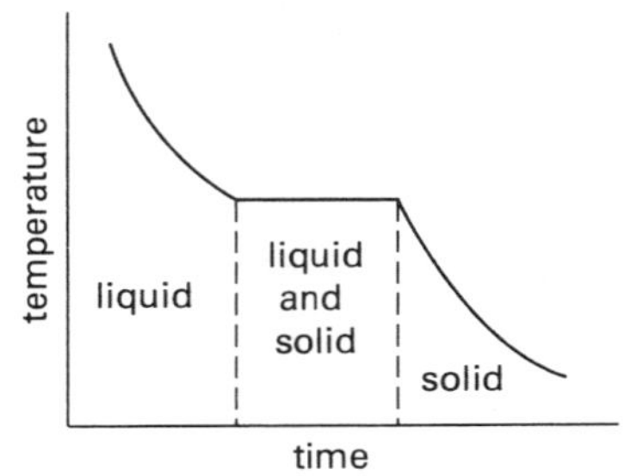

(B)

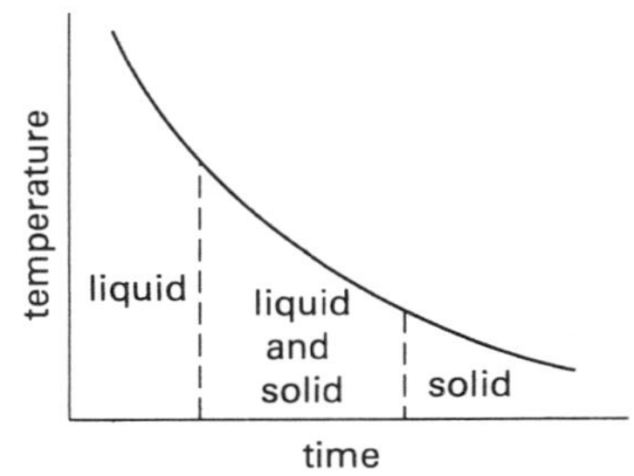

(C)

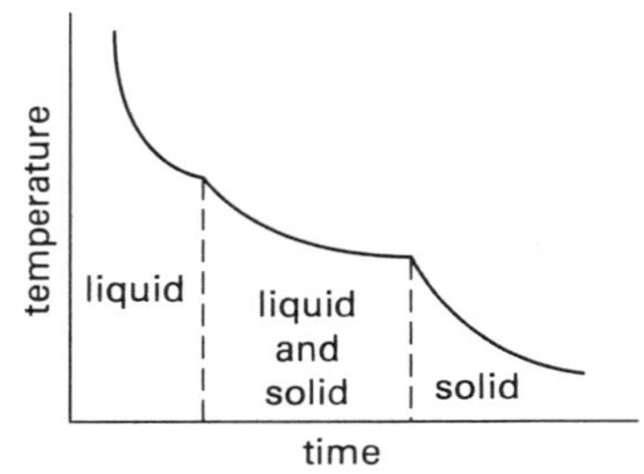

(D)

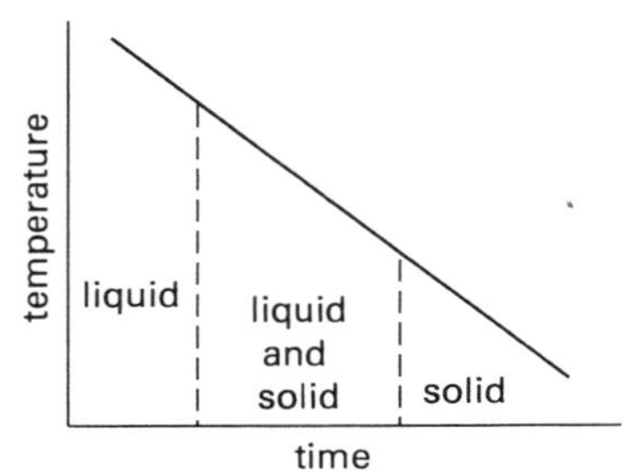

10. Which of the following methods is the most effective in reducing galvanic corrosion between faying objects?

(A) manufacturing both parts from the same material

(B) eliminating moisture in the atmosphere

(C) plating or painting one or both parts with epoxy primer

(D) lubricating the parts

11. Galvanic corrosion between a steel pipe and an attached copper fitting is to be counteracted electrically with a monitored direct current power supply. Most nearly, what should be the applied voltage to eliminate the corrosion?

(A) 0.10 V

(B) 0.34 V

(C) 0.44 V

(D) 0.78 V

12. What metal can be used as a sacrificial anode on a small ocean fishing boat with an aluminum hull?

(A) zinc

(B) magnesium

(C) iron

(D) copper

SOLUTIONS

1. A thermocouple consists of two wires of dissimilar metals that are joined at their two ends. One junction is at a standard, known reference temperature; the other junction is the point whose temperature is being measured. The difference in temperature between the two junctions creates a voltage as a result of the thermoelectric effect. When this voltage and one of the temperatures are known, the temperature difference can be found.

A thermocouple is not a thermometer, and its output represents the ratio of the unknown temperature to the known, standard temperature; it does not represent the absolute temperature.

The answer is (C).

2. Use the phase diagram to find the fraction of solid.

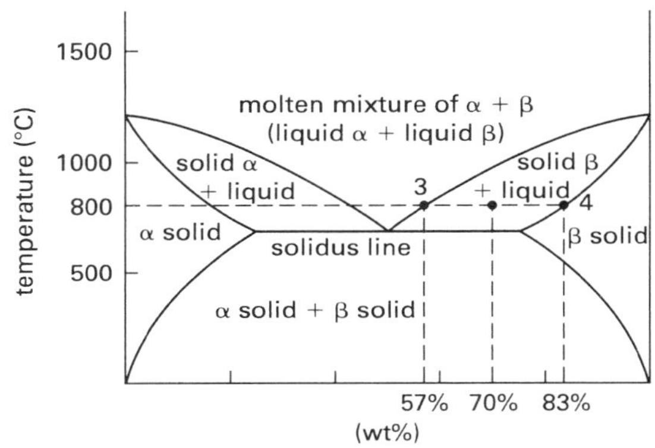

$$\text{wt\% fraction solid} = \frac{x - x_3}{x_4 - x_3} \times 100\%$$

$$= \frac{70\% - 57\%}{83\% - 57\%} \times 100\%$$

$$= 50\%$$

The answer is (D).

3. Martensite is a hard, strong, and brittle material formed by rapid cooling of austenite.

The answer is (D).

4. Steel's hardness, or resistance to deformation, can be increased by surface hardening processes and by some alloying metals. However, tempering and hot working increase the ductility (deformation capability) of steel, not its hardness.

The answer is (D).

5. Hardness in steel is obtained by rapid quenching. Martensite is quenched rapidly, so it has a high hardness (and low ductility).

The answer is (D).

6. Solid and liquid phases are present simultaneously, so the number of phases, P, is 2. Only water is involved, so the number of compounds, C, is 1.

Gibbs phase rule is applicable when both temperature and pressure can be varied. When the temperature is held constant, Gibbs phase rule becomes

$$\begin{aligned} P+F &= C+1\big|_{\text{constant temperature}} \\ F &= C+1-P \\ &= 1+1-2 \\ &= 0 \end{aligned}$$

The answer is (B).

7. The liquidus line divides the diagram into two regions. Above the liquidus line, the alloy is purely liquid, while below the liquidus line, the alloy may exist as solid phase or as a mixture of solid and liquid phases. The liquidus line is CDE.

The answer is (B).

8. The region describes a mixture of solid β component and a liquid of components α and β.

The answer is (D).

9. The solidification of a molten metal is no different than the solidification of water into ice. During the phase change, the temperature remains constant as the heat of fusion is removed. The temperature remains constant during the phase change.

The answer is (A).

10. Faying parts rub against each other. Any paint, plating, coating, or lubricant will be abraded off. Since the parts are already in contact, an electrolyte (moisture) is not needed to complete the circuit. Galvanic action will be reduced if the parts are manufactured from the same material.

The answer is (A).

11. The oxidation potential of the iron in the steel pipe is 0.440 V; the oxidation potential of the copper is −0.337 V. The difference in potentials is $0.440\text{ V} - (-0.337\text{ V}) = 0.777\text{ V}$ (0.78 V).

The answer is (D).

12. Aluminum is near the bottom of the galvanic scale. Only magnesium is lower on the scale, so that magnesium is more anodic and has a greater anode half-cell potential. Therefore, magnesium will become the anode, while the aluminum will become the cathode. The less noble magnesium gives up its electrons to the more noble aluminum. During this process, the magnesium breaks down (corrodes), resulting in magnesium oxide plus free electrons flowing through the electrolyte (sea water) to the aluminum.

The answer is (B).

15 Properties of Materials

PRACTICE PROBLEMS

1. A heating element consists of two wires of different materials connected in series. At 20°C, the two wires have resistances of 600 Ω and 300 Ω and average temperature coefficients of 0.001 1/°C and 0.004 1/°C, respectively. What is most nearly the heating element's total resistance at 50°C?

(A) 900 Ω

(B) 950 Ω

(C) 980 Ω

(D) 990 Ω

2. A solid, cylindrical, copper conductor has a length of 1000 m, a resistance of 0.225 Ω, and a resistivity of 1.77 × 10^{-8} Ω·m. Most nearly, what is the diameter of the conductor?

(A) 1.0 mm

(B) 3.0 mm

(C) 5.0 mm

(D) 10 mm

3. The resistivity of a copper alloy at 20°C is 1.42 × 10^{-8} Ω·m. The thermal coefficient of resistance for the alloy is 0.00402 1/°C. Most nearly, what is the resistivity of the alloy at 100°C?

(A) 1.9 × 10^{-8} Ω·m

(B) 2.0 × 10^{-8} Ω·m

(C) 2.3 × 10^{-8} Ω·m

(D) 1.8 × 10^{-6} Ω·m

4. A rod-shaped electromagnet contains an iron core with a permeability of 86.2 × 10^{-3} H/m. During operation, the magnetic field strength along the centerline of the electromagnet is uniform at 9.76 × 10^{-3} A/m. Most nearly, what is the magnetic flux density in the iron core?

(A) 4.2 × 10^{-4} Wb/m^2

(B) 8.4 × 10^{-4} Wb/m^2

(C) 1.1 × 10^{-1} Wb/m^2

(D) 8.8 Wb/m^2

5. A capacitor is constructed from two circular plates 1 cm in diameter. The plates are parallel and separated by 2 mm of a dielectric. The capacitor's capacitance is 35 pF. Most nearly, what is the permittivity of the dielectric?

(A) 6.5 × 10^{-11} F/m

(B) 8.9 × 10^{-11} F/m

(C) 7.0 × 10^{-10} F/m

(D) 8.9 × 10^{-10} F/m

6. A certain kind of test might result in the stress-strain diagram shown.

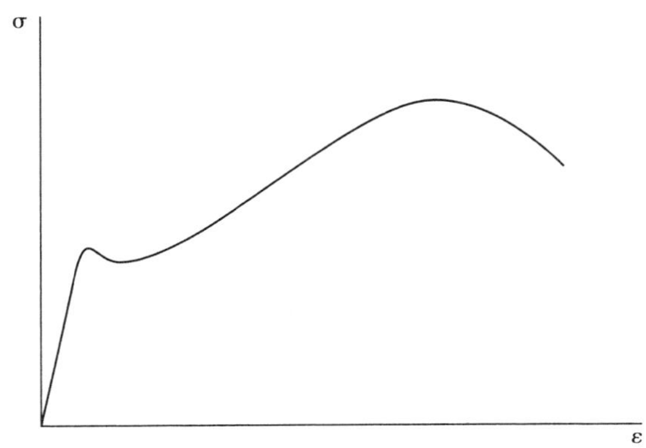

What kind of test might result from this diagram?

(A) resilience test

(B) rotating beam test

(C) ductility test

(D) tensile test

7. What term is used for the ratio of stress to strain below the proportional limit?

(A) modulus of rigidity

(B) Hooke's constant

(C) Poisson's ratio

(D) Young's modulus

8. What does an impact test measure?

(A) hardness

(B) yield strength

(C) toughness

(D) creep strength

9. A stress-strain diagram is shown.

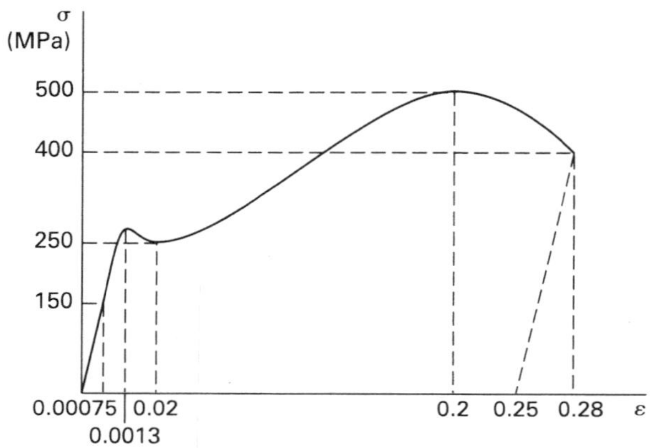

Most nearly, what is the percent elongation at failure?

(A) 14%

(B) 19%

(C) 25%

(D) 28%

10. In this illustration, what does the value of 40 MPa represent?

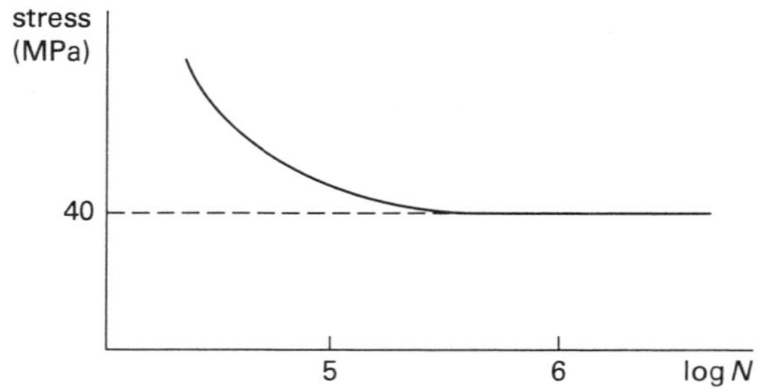

I. fatigue limit

II. endurance limit

III. proportional limit

IV. yield stress

(A) I and II

(B) I and IV

(C) II and III

(D) III and IV

11. If δ is deformation, and L is the original length of the specimen, what is the definition of normal strain, ε?

(A) $\varepsilon = \dfrac{L+\delta}{L}$

(B) $\varepsilon = \dfrac{L+\delta}{\delta}$

(C) $\varepsilon = \dfrac{\delta}{L+\delta}$

(D) $\varepsilon = \dfrac{\delta}{L}$

12. An electrochemical cell is shown.

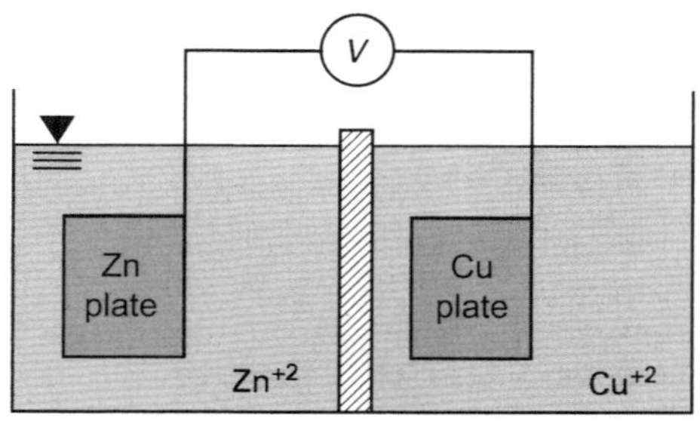

What is the reaction at the anode?

(A) $Cu \rightarrow Cu^{2+} + 2e^-$

(B) $Cu^{2+} + 2e^- \rightarrow Cu$

(C) $Zn \rightarrow Zn^{2+} + 2e^-$

(D) $Zn^{2+} + 2e^- \rightarrow Zn$

13. Corrosion of iron can be inhibited with a more electropositive coating, while a less electropositive coating tends to accelerate corrosion. Which of these coatings will contribute to corrosion of iron products?

(A) zinc

(B) gold

(C) aluminum

(D) magnesium

SOLUTIONS

1. Find the resistance of each wire at 50°C.

For wire A,

$$\begin{aligned} R_A &= R_0[1 + \alpha(T - T_0)] \\ &= (600\ \Omega)\left(1 + \left(0.001\ \frac{1}{°C}\right)(50°C - 20°C)\right) \\ &= 618\ \Omega \end{aligned}$$

For wire B,

$$\begin{aligned} R_B &= R_0[1 + \alpha(T - T_0)] \\ &= (300\ \Omega)\left(1 + \left(0.004\ \frac{1}{°C}\right)(50°C - 20°C)\right) \\ &= 336\ \Omega \end{aligned}$$

The total resistance of the heating element is

$$\begin{aligned} R_{total} &= R_A + R_B \\ &= 618\ \Omega + 336\ \Omega \\ &= 954\ \Omega \quad (950\ \Omega) \end{aligned}$$

The answer is (B).

2. Use the formula for resistance to find the cross-sectional area of the conductor.

$$\begin{aligned} R &= \frac{\rho L}{A} \\ A &= \frac{\rho L}{R} = \frac{(1.77 \times 10^{-8}\ \Omega\cdot\text{m})(1000\ \text{m})}{0.225\ \Omega} \\ &= 7.87 \times 10^{-5}\ \text{m}^2 \end{aligned}$$

Find the diameter of the conductor from its cross-sectional area.

$$\begin{aligned} A &= \frac{\pi d^2}{4} \\ d &= \sqrt{\frac{4A}{\pi}} = \sqrt{\frac{(4)(7.87 \times 10^{-5}\ \text{m}^2)}{\pi}}\left(1000\ \frac{\text{mm}}{\text{m}}\right) \\ &= 10.01\ \text{mm} \quad (10\ \text{mm}) \end{aligned}$$

The answer is (D).

Properties of Electrical Mats.

3. Use the formula that relates the resistivity to temperature.

$$\begin{aligned}\rho &= \rho_0[1+\alpha(T-T_0)]\\ &= (1.42\times10^{-8}\ \Omega\cdot\text{m})\left(1+\left(0.00402\ \frac{1}{^\circ\text{C}}\right)(100^\circ\text{C}-20^\circ\text{C})\right)\\ &= 1.88\times10^{-8}\ \Omega\cdot\text{m}\quad(1.9\times10^{-8}\ \Omega\cdot\text{m})\end{aligned}$$

The answer is (A).

4. Use the equation for magnetic field strength and solve for the magnetic flux density.

$$\mathbf{H} = \frac{\mathbf{B}}{\mu}$$

$$\mathbf{B} = \mu\mathbf{H}$$

$$\begin{aligned}|\mathbf{B}| = \mu|\mathbf{H}| &= \left(9.76\times10^{-3}\ \frac{\text{A}}{\text{m}}\right)\left(86.2\times10^{-3}\ \frac{\text{H}}{\text{m}}\right)\\ &= 8.41\times10^{-4}\ \text{Wb/m}^2\quad(8.4\times10^{-4}\ \text{Wb/m}^2)\end{aligned}$$

The answer is (B).

5. The area of each plate is

$$\begin{aligned}A &= \frac{\pi D^2}{4}\\ &= \frac{\pi(1\times10^{-2}\ \text{m})^2}{4}\\ &= 7.85\times10^{-5}\ \text{m}^2\end{aligned}$$

Use the formula for capacitance and solve for permittivity.

$$C = \frac{\varepsilon A}{d}$$

$$\begin{aligned}\varepsilon &= \frac{Cd}{A}\\ &= \frac{(35\times10^{-12}\ \text{F})(2\times10^{-3}\ \text{m})}{7.85\times10^{-5}\ \text{m}^2}\\ &= 8.92\times10^{-10}\ \text{F/m}\quad(8.9\times10^{-10}\ \text{F/m})\end{aligned}$$

The answer is (D).

6. The diagram shows the results from a tensile test.

Both resilience and ductility may be calculated from the results, but the test is not known by those names. The rotating beam is a cyclic test and does not yield a monotonic stress-strain curve.

The answer is (D).

7. Young's modulus is defined by Hooke's law.

$$\sigma = E\varepsilon$$

E is Young's modulus, also called the modulus of elasticity, and is equal to stress divided by strain within the proportional region of the stress-strain curve.

The answer is (D).

8. An impact test measures the energy needed to fracture a test sample. This is a measure of toughness.

The answer is (C).

9. The strain at failure used in the equation is found by extending a line from the failure point to the strain axis, parallel to the linear portion of the curve. The percent elongation is an indicator of the ductility of a material, but it is not the same as the ductility. The percent elongation is

$$\begin{aligned}\text{percent elongation} &= \varepsilon_f\times100\%\\ &= 0.25\times100\%\\ &= 25\%\end{aligned}$$

The answer is (C).

10. The illustration shows results of an endurance (or fatigue) test. The value of 40 MPa is called the endurance stress, endurance limit, or fatigue limit, and is equal to the maximum stress that can be repeated indefinitely without causing the specimen to fail.

The answer is (A).

11. Strain is defined as elongation, δ, per unit length, L.

The answer is (D).

12. Zinc has a higher potential and will act as the anode. By definition, the anode is where electrons are lost. The reaction at the anode of the electrochemical cell is $Zn \rightarrow Zn^{2+} + 2e^-$.

The answer is (C).

13. Zinc, aluminum, and magnesium are all more electropositive (anodic) than iron and will corrode sacrificially to protect it. Gold is more cathodic and will be protected at the expense of the iron.

The answer is (B).

16 Properties of Semiconductor Materials

PRACTICE PROBLEMS

1. A 9 V battery is connected across a semiconductor with a resistive diffused layer 1 cm wide, 10 cm long, and 1 cm thick. The diffused layer is doped with an n-type dopant to a concentration of 8.34×10^{17} carriers/cm^3. The intrinsic carrier concentration in the diffused layer is 10^{10} carriers/cm^3, the mobility of holes is 150 cm^2/V·s, and the electron mobility is 360 cm^2/V·s. Most nearly, what is the current through the diffused layer?

(A) 0.12 A

(B) 18 A

(C) 43 A

(D) 61 A

2. When placed in a particular electric field, an n-type semiconductor experiences a drift current of 1 kA/cm^2. The semiconductor is doped to a concentration of 8.76×10^{14} carriers/cm^3. The intrinsic carrier concentration is 10^{10} carriers/cm^3, the mobility of holes is 490 cm^2/V·s, and the mobility of electrons is 1310 cm^2/V·s. Most nearly, what is the strength of the electric field?

(A) 2.7 kV/cm

(B) 5.4 kV/cm

(C) 8.5 kV/cm

(D) 15 kV/cm

3. A p-type semiconductor is doped to a concentration of 5.65×10^{15} carriers/cm^3. The intrinsic carrier concentration of the semiconductor is 10^{10} carriers/cm^3, the mobility of holes is 460 cm^2/V·s, and the mobility of electrons is 1260 cm^2/V·s. The total drift current density, J, produced in a semiconductor is equal to the semiconductor's conductivity, σ, multiplied by the electric field applied to the semiconductor, E.

$$J = \sigma E$$

The semiconductor is placed in an electric field with a strength of 75 V/cm. Most nearly, what is the drift current density produced?

(A) 0.19 A/cm^2

(B) 31 A/cm^2

(C) 85 A/cm^2

(D) 120 A/cm^2

4. A germanium-based n-type semiconductor is doped such that the negative charge carrier density far exceeds the positive charge carrier density (i.e., $n \gg p$). The conductivity of the semiconductor is 0.643 S/cm, the intrinsic carrier concentration is 10^{10} carriers/cm^3, the mobility of holes is 2970 cm^2/V·s, and the mobility of electrons is 8860 cm^2/V·s. Most nearly, the electron carrier concentration in the semiconductor is

(A) 1.5×10^{11} carriers/cm^3

(B) 2.8×10^{14} carriers/cm^3

(C) 4.5×10^{14} carriers/cm^3

(D) 1.4×10^{15} carriers/cm^3

5. An n-type semiconductor has a dopant concentration of 4.53×10^{15} carriers/cm^3, an intrinsic carrier concentration of 10^{10} carriers/cm^3, a mobility of holes of 460 cm^2/V·s, and a mobility of electrons of 1260 cm^2/V·s. Most nearly, the resistivity of the semiconductor is

(A) 0.91 Ω·cm

(B) 1.1 Ω·cm

(C) 3.0 Ω·cm

(D) 1400 Ω·cm

SOLUTIONS

1. Use the law of mass action to find the hole concentration.

$$(p)(n) = n_i^2$$

$$p = \frac{n_i^2}{n} = \frac{\left(1.0 \times 10^{10}\ \frac{\text{carriers}}{\text{cm}^3}\right)^2}{8.34 \times 10^{17}\ \frac{\text{carriers}}{\text{cm}^3}}$$
$$= 1.20 \times 10^2\ \text{carriers/cm}^3$$

The conductivity of the semiconductor is

$$\sigma = q(n\mu_n + p\mu_p)$$
$$= (1.6 \times 10^{-19}\ \text{C})\left(\begin{array}{l}\left(8.34 \times 10^{17}\ \frac{\text{carriers}}{\text{cm}^3}\right)\left(360\ \frac{\text{cm}^2}{\text{V·s}}\right) \\ +\left(1.20 \times 10^2\ \frac{\text{carriers}}{\text{cm}^3}\right)\left(150\ \frac{\text{cm}^2}{\text{V·s}}\right)\end{array}\right)$$
$$= 48.04\ \text{S/cm}$$

Resistivity is the reciprocal of conductivity.

$$\rho = \frac{1}{\sigma} = \frac{1}{48.04\ \frac{\text{S}}{\text{cm}}}$$
$$= 0.0208\ \Omega\text{·cm}$$

The sheet resistance is

$$R_s = \frac{\rho}{d} = \frac{0.0208\ \Omega\text{·cm}}{1\ \text{cm}}$$
$$= 0.0208\ \Omega \quad \text{(per square)}$$

The resistance of the diffused layer is

$$R = R_s\left(\frac{L}{W}\right) = (0.0208\ \Omega)\left(\frac{10\ \text{cm}}{1\ \text{cm}}\right)$$
$$= 0.208\ \Omega$$

Use Ohm's law to find the current through the diffused layer.

$$V = IR$$
$$I = \frac{V}{R} = \frac{9\ \text{V}}{0.208\ \Omega}$$
$$= 43.3\ \text{A} \quad (43\ \text{A})$$

The answer is (C).

2. Use the law of mass action to find the hole concentration.

$$(p)(n) = n_i^2$$

$$p = \frac{n_i^2}{n} = \frac{\left(1.0 \times 10^{10}\ \frac{\text{carriers}}{\text{cm}^3}\right)^2}{8.76 \times 10^{14}\ \frac{\text{carriers}}{\text{cm}^3}}$$
$$= 1.14 \times 10^5\ \text{carriers/cm}^3$$

The conductivity of the semiconductor is

$$\sigma = q(n\mu_n + p\mu_p)$$
$$= (1.6 \times 10^{-19}\ \text{C})\left(\begin{array}{l}\left(8.76 \times 10^{14}\ \frac{\text{carriers}}{\text{cm}^3}\right)\left(1310\ \frac{\text{cm}^2}{\text{V·s}}\right) \\ +\left(1.14 \times 10^5\ \frac{\text{carriers}}{\text{cm}^3}\right)\left(490\ \frac{\text{cm}^2}{\text{V·s}}\right)\end{array}\right)$$
$$= 0.184\ \text{S/cm}$$

Use the equation for drift current density to find the strength of the electric field.

$$J = \sigma E$$

$$E = \frac{J}{\sigma} = \frac{1\ \frac{\text{kA}}{\text{cm}^2}}{0.184\ \frac{\text{S}}{\text{cm}}}$$
$$= 5.44\ \text{kV/cm} \quad (5.4\ \text{kV/cm})$$

The answer is (B).

3. Use the law of mass action to find the hole concentration.

$$(p)(n) = n_i^2$$

$$p = \frac{n_i^2}{n} = \frac{\left(1.0 \times 10^{10}\ \frac{\text{carriers}}{\text{cm}^3}\right)^2}{5.65 \times 10^{15}\ \frac{\text{carriers}}{\text{cm}^3}}$$
$$= 1.77 \times 10^4\ \text{carriers/cm}^3$$

The conductivity of the semiconductor is

$$\sigma = q(n\mu_n + p\mu_p)$$
$$= (1.6\times10^{-19}\ \text{C})\left(\begin{array}{l}\left(5.65\times10^{15}\ \dfrac{\text{carriers}}{\text{cm}^3}\right)\left(1260\ \dfrac{\text{cm}^2}{\text{V·s}}\right)\\ +\left(1.77\times10^{4}\ \dfrac{\text{carriers}}{\text{cm}^3}\right)\left(460\ \dfrac{\text{cm}^2}{\text{V·s}}\right)\end{array}\right)$$
$$= 1.139\ \text{S/cm}$$

When the semiconductor is placed in the electric field, the drift current density will be

$$J = \sigma E = \left(1.139\ \frac{\text{S}}{\text{cm}}\right)\left(75\ \frac{\text{V}}{\text{cm}}\right)$$
$$= 85.4\ \text{A/cm}^2\quad (85\ \text{A/cm}^2)$$

The answer is (C).

4. The conductivity of the semiconductor is

$$\sigma = q(n\mu_n + p\mu_p)$$

Since $n \gg p$, consider the $p\mu_p$ term to be zero. Rearrange the simplified equation to solve for the electron carrier concentration.

$$\sigma \approx qn\mu_n \quad [n \gg p]$$
$$n \approx \frac{\sigma}{q\mu_n} = \frac{0.643\ \dfrac{\text{S}}{\text{cm}}}{(1.6\times10^{-19}\ \text{C})\left(8860\ \dfrac{\text{cm}^2}{\text{V·s}}\right)}$$
$$= 4.53\times10^{14}\ \text{carriers/cm}^3\quad (4.5\times10^{14}\ \text{carriers/cm}^3)$$

The answer is (C).

5. Use the law of mass action to find the hole concentration.

$$(p)(n) = n_i^2$$
$$p = \frac{n_i^2}{n} = \frac{\left(1.0\times10^{10}\ \dfrac{\text{carriers}}{\text{cm}^3}\right)^2}{4.53\times10^{15}\ \dfrac{\text{carriers}}{\text{cm}^3}}$$
$$= 2.21\times10^{4}\ \text{carriers/cm}^3$$

The conductivity of the semiconductor is

$$\sigma = q(n\mu_n + p\mu_p)$$
$$= (1.6\times10^{-19}\ \text{C})\left(\begin{array}{l}\left(4.53\times10^{15}\ \dfrac{\text{carriers}}{\text{cm}^3}\right)\left(1260\ \dfrac{\text{cm}^2}{\text{V·s}}\right)\\ +\left(2.21\times10^{4}\ \dfrac{\text{carriers}}{\text{cm}^3}\right)\left(460\ \dfrac{\text{cm}^2}{\text{V·s}}\right)\end{array}\right)$$
$$= 0.9132\ \text{S/cm}$$

The resistivity is the reciprocal of conductivity.

$$\rho = \frac{1}{\sigma} = \frac{1}{0.9132\ \dfrac{\text{S}}{\text{cm}}}$$
$$= 1.095\ \Omega\text{·cm}\quad (1.1\ \Omega\text{·cm})$$

The answer is (B).

17 Properties of Electrical Devices and Circuits

PRACTICE PROBLEMS

1. Four identical capacitors are assembled into a circuit as shown. Most nearly, what is the total capacitance between terminals A and B?

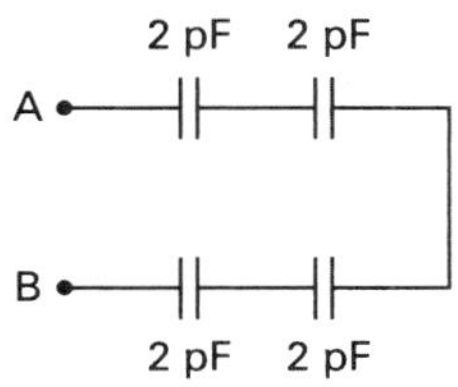

(A) 0.5 pF

(B) 2 pF

(C) 4 pF

(D) 6 pF

2. The circuit shown is at steady state, and all circuit elements are ideal.

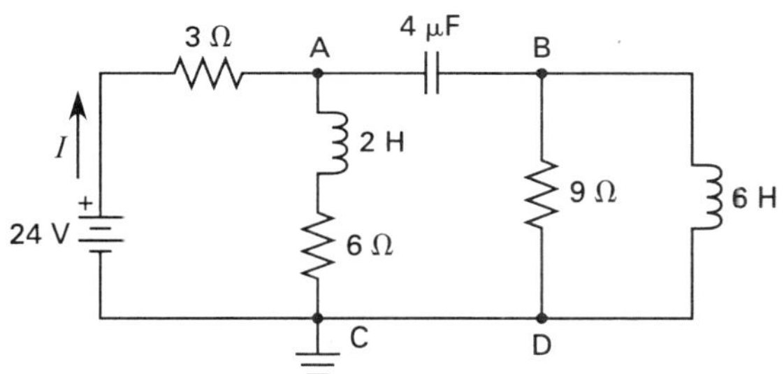

Most nearly, what is the energy stored in the 2 H inductor?

(A) 7.1 J

(B) 13 J

(C) 14 J

(D) 29 J

3. An ideal battery and three resistors are assembled in a circuit as shown. Most nearly, what is the current flowing through the battery?

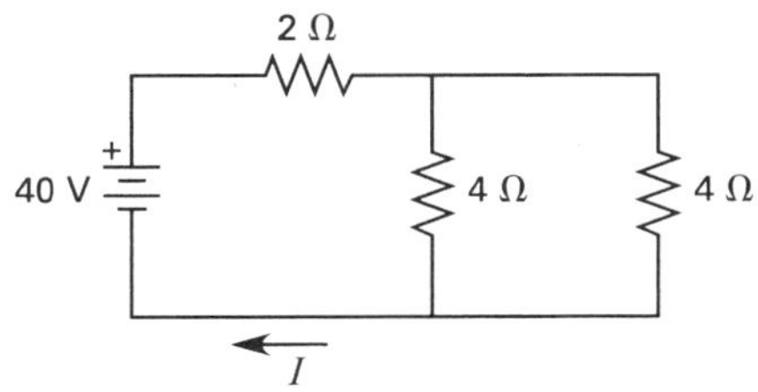

(A) 4.0 A

(B) 6.7 A

(C) 10 A

(D) 15 A

4. In the circuit shown, each resistor has a resistance of 10 Ω.

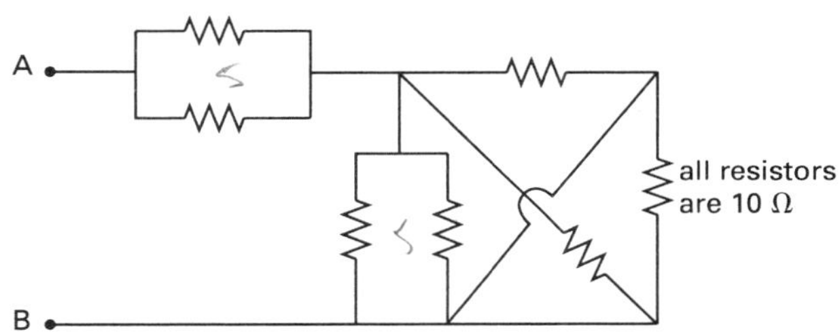

Most nearly, what is the total resistance between terminals A and B?

(A) 7.5 Ω

(B) 7.9 Ω

(C) 8.3 Ω

(D) 10 Ω

5. A charge of 0.75 C passes through a wire every 15 s. Most nearly, what is the current in the wire?

(A) 5.0 mA

(B) 10 mA

(C) 20 mA

(D) 50 mA

6. A current of 10 A flows through a 1 mm diameter wire. Most nearly, what is the average number of electrons per second that pass through a cross section of the wire?

(A) 1.6×10^{18} electrons/s

(B) 6.2×10^{18} electrons/s

(C) 1.6×10^{19} electrons/s

(D) 6.3×10^{19} electrons/s

7. The circuit shown is in steady state.

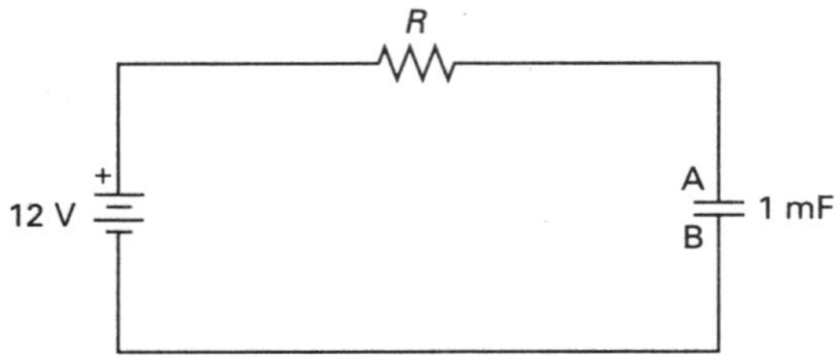

What is most nearly the charge on the capacitor on plate A?

(A) 83 pC

(B) 120 pC

(C) 83 μC

(D) 0.012 C

8. Two parallel plates with a potential difference of 100 V generate an electric field of 100×10^3 V/m between them. Most nearly, what is the distance between the plates?

(A) 1.0×10^{-3} m

(B) 2.0×10^{-3} m

(C) 2.0×10^{-2} m

(D) 5.0×10^{-2} m

9. A resistive network is constructed from four ideal resistors.

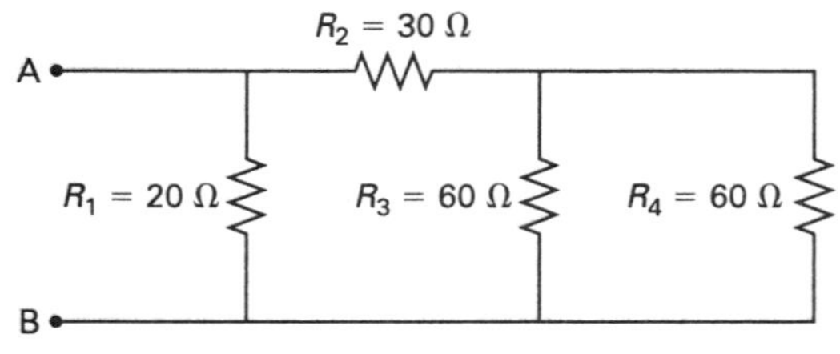

Most nearly, what is the equivalent resistance between terminals A and B?

(A) 5.0 Ω

(B) 10 Ω

(C) 12 Ω

(D) 15 Ω

10. A 50 μF capacitor starts out with zero voltage across it. When a switch is closed at $t = 0$, the capacitor passes a current that varies with time according to the function shown.

$$i_C(t) = (0.01 \text{ A})e^{-t/0.05 \text{ s}}$$

Most nearly, what is the energy stored in the capacitor 0.1 s after the switch is closed?

(A) 4.6×10^{-5} J

(B) 1.9×10^{-3} J

(C) 2.4×10^{-3} J

(D) 3.5×10^{-3} J

11. An inductor with an inductance of 0.01 H is made up of a core and coil of 100 turns. The wire is 1×10^{-2} m long, and the cross-sectional area of the inductor is 3×10^{-5} m^2. Most nearly, what is the relative permeability of the inductor core?

(A) 110

(B) 160

(C) 210

(D) 270

12. A circuit is constructed of four capacitors as shown.

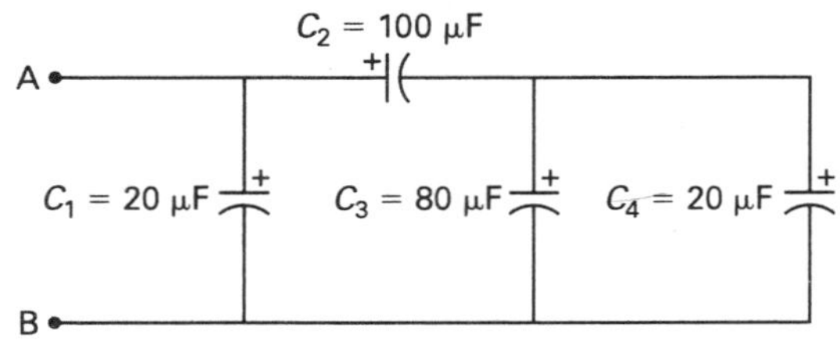

Most nearly, what is the equivalent capacitance between terminals A and B?

(A) 20 μF

(B) 50 μF

(C) 70 μF

(D) 100 μF

13. A circuit is constructed from four inductors as shown.

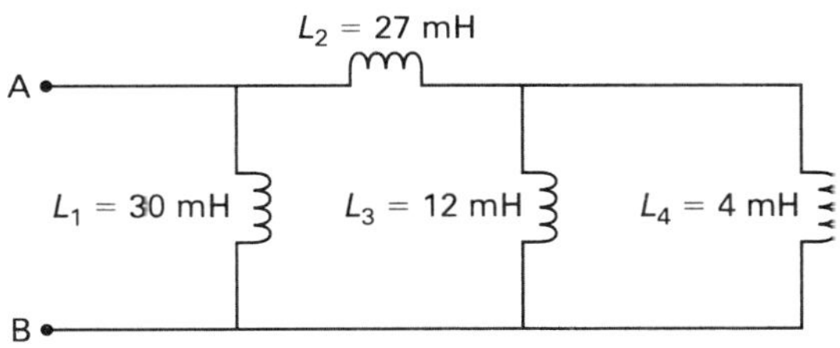

(A) 8 mH

(B) 10 mH

(C) 12 mH

(D) 15 mH

SOLUTIONS

1. Use the equation for capacitors connected in series.

$$
\begin{aligned}
C_S &= \frac{1}{1/C_1 + 1/C_2 + \cdots + 1/C_n} \\
&= \frac{1}{\dfrac{1}{2\text{ pF}} + \dfrac{1}{2\text{ pF}} + \dfrac{1}{2\text{ pF}} + \dfrac{1}{2\text{ pF}}} \\
&= 0.5\text{ pF}
\end{aligned}
$$

The answer is (A).

2. A capacitor in a DC circuit has an infinite resistance. There is no current through the capacitor between nodes A and B. An inductor in a DC circuit has no resistance, so the circuit simplifies to

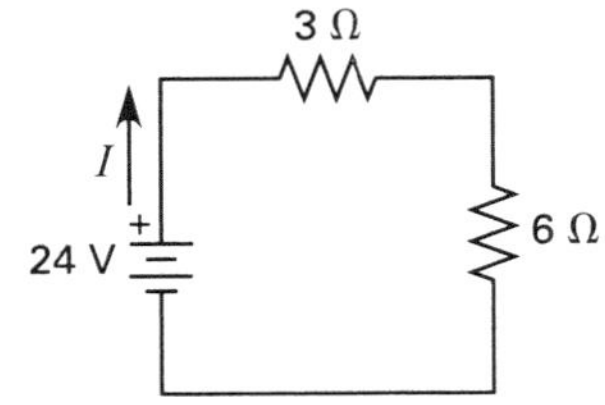

Use Ohm's law to find the current.

$$
\begin{aligned}
V &= IR \\
I &= \frac{V}{R} = \frac{24\text{ V}}{3\ \Omega + 6\ \Omega} \\
&= 2.67\text{ A}
\end{aligned}
$$

The entire current passes through the 2 H inductor. The energy stored in the inductor is

$$
\begin{aligned}
\text{energy} &= Li_L^2/2 \\
&= \frac{(2\text{ H})(2.67\text{ A})^2}{2} \\
&= 7.13\text{ J} \quad (7.1\text{ J})
\end{aligned}
$$

The answer is (A).

3. Simplify the circuit. Find the combined resistance of the two 4 Ω resistors in parallel.

$$
\begin{aligned}
R_P &= 1/(1/R_1 + 1/R_2 + \cdots + R_n) \\
&= \frac{1}{\dfrac{1}{4\ \Omega} + \dfrac{1}{4\ \Omega}} \\
&= 2\ \Omega
\end{aligned}
$$

The total resistance of the circuit is the combined resistance of the 2 Ω resistor and R^P in series.

$$\begin{aligned} R_S &= R_1 + R_2 + \cdots + R_n \\ &= 2\ \Omega + 2\ \Omega \\ &= 4\ \Omega \end{aligned}$$

Use Ohm's law to find the current through the battery.

$$\begin{aligned} V &= IR \\ I &= \frac{V}{R} = \frac{40\ \text{V}}{4\ \Omega} \\ &= 10\ \text{A} \end{aligned}$$

The answer is (C).

4. Untwist the network. Both terminals of the rightmost resistor are connected to the same point, and that resistor is omitted.

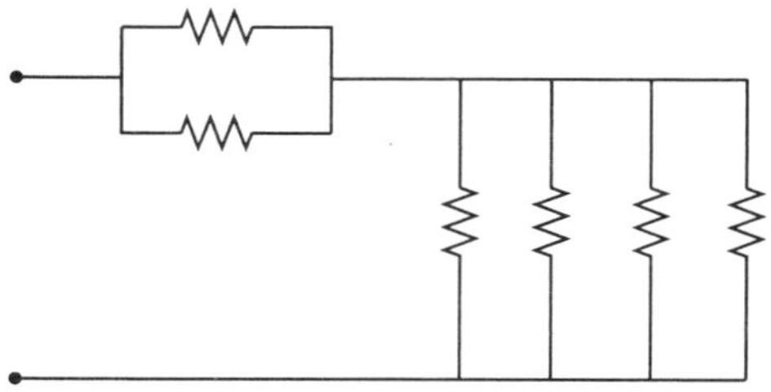

The circuit consists of a set of two parallel resistors in series with a set of four parallel resistors. The resistance of the first set is

$$R_{P,1} = \frac{R_1R_2}{R_1 + R_2} = \frac{(10\ \Omega)(10\ \Omega)}{10\ \Omega + 10\ \Omega} = 5\ \Omega$$

The resistance of the second set is

$$\begin{aligned} R_{P,2} &= 1/(1/R_1 + 1/R_2 + \cdots + R_n) \\ &= \frac{1}{\frac{1}{10\ \Omega} + \frac{1}{10\ \Omega} + \frac{1}{10\ \Omega} + \frac{1}{10\ \Omega}} \\ &= 2.5\ \Omega \end{aligned}$$

The resistance of the two sets in series is

$$\begin{aligned} R_S &= R_1 + R_2 + \cdots + R_n \\ &= R_{P,1} + R_{P,2} \\ &= 5\ \Omega + 2.5\ \Omega \\ &= 7.5\ \Omega \end{aligned}$$

The answer is (A).

5. Current is the charge per unit time passing through the wire.

$$I = \frac{q}{t} = \frac{(0.75\ \text{C})\left(1000\ \frac{\text{mA}}{\text{A}}\right)}{15\ \text{s}} = 50\ \text{mA}$$

The answer is (D).

6. A current of 10 A is equivalent to 10 C/s. One electron has a charge of approximately 1.6×10^{-19} C.

$$\begin{aligned} \dot{q} &= \frac{I}{Q} = \frac{10\ \frac{\text{C}}{\text{s}}}{1.6 \times 10^{-19}\ \frac{\text{C}}{\text{electron}}} \\ &= 6.25 \times 10^{19}\ \text{electrons/s} \quad (6.3 \times 10^{19}\ \text{electrons/s}) \end{aligned}$$

The wire diameter is irrelevant.

The answer is (D).

7. In steady state, all of the voltage is across the capacitor.

$$\begin{aligned} Q &= CV = (1 \times 10^{-3}\ \text{F})(12\ \text{V}) \\ &= 0.012\ \text{C} \end{aligned}$$

The answer is (D).

8. Use the equation for electric field strength, and solve for the distance between the plates.

$$\begin{aligned} E &= \frac{V}{d} \\ d &= \frac{V}{E} = \frac{100\ \text{V}}{100 \times 10^3\ \frac{\text{V}}{\text{m}}} \\ &= 1.0 \times 10^{-3}\ \text{m} \end{aligned}$$

The answer is (A).

9. Find the equivalent resistance of resistor 3 and resistor 4 in parallel.

$$\begin{aligned} R_P &= \frac{R_1R_2}{R_1 + R_2} \\ R_{34} &= \frac{R_3R_4}{R_3 + R_4} = \frac{(60\ \Omega)(60\ \Omega)}{60\ \Omega + 60\ \Omega} \\ &= 30\ \Omega \end{aligned}$$

Properties of Electrical Mats.

This equivalent resistance, R_{34}, is in series with resistor 2. The equivalent resistance of resistor 2 and R_{34} in series is

$$R_S = R_1 + R_2 + \cdots + R_n$$
$$R_{234} = R_2 + R_{34} = 30\ \Omega + 30\ \Omega = 60\ \Omega$$

The total resistance between terminals A and B is equivalent to R_{234} in parallel with resistor 1. The equivalent resistance of resistor 1 and R_{234} in parallel is

$$R_{\text{AE}} = \frac{R_1 R_{234}}{R_1 + R_{234}} = \frac{(20\ \Omega)(60\ \Omega)}{20\ \Omega + 60\ \Omega} = 15\ \Omega$$

The answer is (D).

10. The voltage across the capacitor at $t = 0.1$ s can be obtained by integration.

$$\begin{aligned} v_C(t) &= v_C(0) + \frac{1}{C}\int_0^t i_C(\tau)\,d\tau \\ &= 0 + \frac{1}{50 \times 10^{-6}\ \text{F}}\int_0^{0.1\ \text{s}} (0.01\ \text{A})e^{-\tau/0.05\ \text{s}}\,d\tau \\ &= \left(\frac{(0.01\ \text{A})(-0.05\ \text{s})}{50 \times 10^{-6}\ \text{F}}\right)e^{-\tau/0.05\ \text{s}}\Bigg|_{0\ \text{s}}^{0.1\ \text{s}} \\ &= -1.353\ \text{V} - (-10\ \text{V}) \\ &= 8.65\ \text{V} \end{aligned}$$

The energy in the capacitor is

$$\text{energy} = Cv_C^2/2 = \frac{(50 \times 10^{-6}\ \text{F})(8.65\ \text{V})^2}{2} = 1.87 \times 10^{-3}\ \text{J} \quad (1.9 \times 10^{-3}\ \text{J})$$

The answer is (B).

11. The permeability of the inductor core is the product of the core's relative permeability and the permeability of free space, $4\pi \times 10^{-7}$ H/m.

$$\mu = \kappa_m \mu_0$$

Rearrange the equations for the inductance of a coil to solve for the relative permeability of the core.

$$L = N^2\mu A/l = \frac{N^2(\kappa_m \mu_0)A}{l}$$

$$\kappa_m = \frac{Ll}{N^2\mu_0 A} = \frac{(0.01\ \text{H})(1 \times 10^{-2}\ \text{m})}{(100)^2\left(4\pi \times 10^{-7}\ \frac{\text{H}}{\text{m}}\right)(3 \times 10^{-5}\ \text{m}^2)} = 265.27 \quad (270)$$

The answer is (D).

12. Find the equivalent capacitance of capacitor 3 and capacitor 4 in parallel.

$$C_P = C_1 + C_2 + \cdots + C_n$$
$$C_{34} = C_3 + C_4 = 80\ \mu\text{F} + 20\ \mu\text{F} = 100\ \mu\text{F}$$

This equivalent capacitance, C_{34}, is in series with capacitor 2. The equivalent capacitance of capacitor 2 and C_{34} in series is

$$C_S = \frac{1}{1/C_1 + 1/C_2 + \cdots + 1/C_n}$$
$$C_{234} = \frac{1}{\frac{1}{C_2} + \frac{1}{C_{34}}} = \frac{1}{\frac{1}{100\ \mu\text{F}} + \frac{1}{100\ \mu\text{F}}} = 50\ \mu\text{F}$$

The total capacitance between terminals A and B is equivalent to C_{234} in parallel with capacitor 1. The equivalent capacitance of capacitor 1 and C_{234} in parallel is

$$C_{\text{AB}} = C_1 + C_{234} = 20\ \mu\text{F} + 50\ \mu\text{F} = 70\ \mu\text{F}$$

The answer is (C).

13. Find the equivalent inductance of inductor 3 and inductor 4 in parallel.

$$L_P = \frac{1}{1/L_1 + 1/L_2 + \cdots + 1/L_n}$$
$$L_{34} = \frac{1}{\frac{1}{L_3} + \frac{1}{L_4}} = \frac{1}{\frac{1}{12\ \text{mH}} + \frac{1}{4\ \text{mH}}} = 3\ \text{mH}$$

Properties of Electrical Mats.

The equivalent inductance, L_{34}, is in series with inductor 2. The equivalent inductance of inductor 2 and L_{34} in series is

$$\begin{aligned} L_S &= L_1 + L_2 + \cdots + L_n \\ L_{234} &= L_2 + L_{34} = 27 \text{ mH} + 3 \text{ mH} \\ &= 30 \text{ mH} \end{aligned}$$

The total inductance between terminals A and B is equivalent to L_{234} in parallel with inductor 1. The equivalent inductance of inductor 1 and L_{234} in parallel is

$$\begin{aligned} L_P &= \frac{1}{1/L_1 + 1/L_2 + \cdots + 1/L_n} \\ L_{\text{AB}} &= \frac{1}{\frac{1}{L_1} + \frac{1}{L_{234}}} = \frac{1}{\frac{1}{30 \text{ mH}} + \frac{1}{30 \text{ mH}}} \\ &= 15 \text{ mH} \end{aligned}$$

The answer is (D).

18 Energy, Work, and Power

PRACTICE PROBLEMS

1. A bullet of mass 100 g is fired at a wooden block resting on a frictionless, horizontal surface as shown. A spring with a stiffness of 53 kN/cm resists the motion of the block. The maximum displacement of the block produced by the impact of the bullet is 3.4 cm. There are no losses at impact, and the spring has no mass.

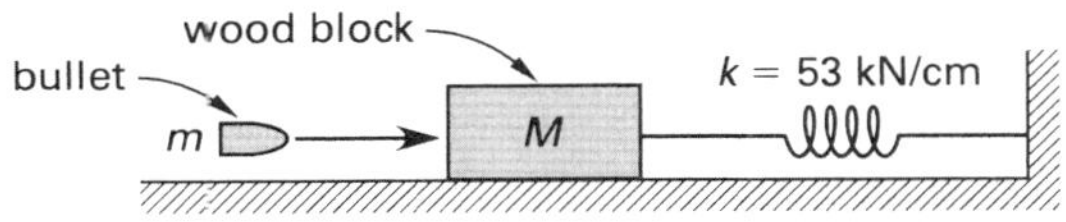

Most nearly, the velocity of the bullet at impact is

(A) 250 km/h

(B) 450 km/h

(C) 630 km/h

(D) 890 km/h

2. A 0.05 kg mass attached to a spring is accelerated to a velocity of 0.4 m/s over a distance of 0.1 m. The spring's spring constant is 0.5 N/m. The spring's mass is negligible.

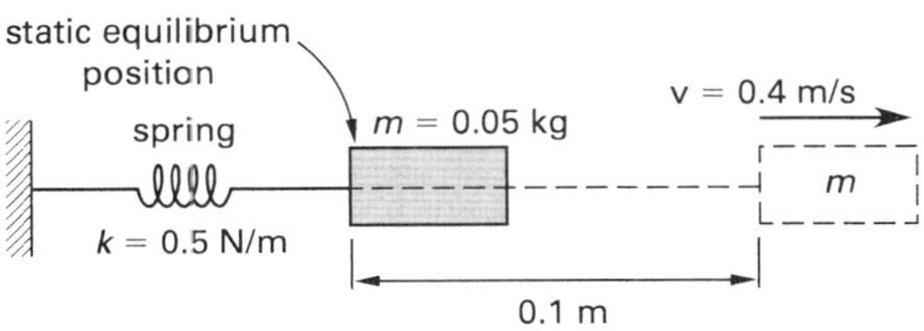

Most nearly, what is the total energy of the mass?

(A) 0.0025 J

(B) 0.0040 J

(C) 0.0065 J

(D) 0.0092 J

3. In the mass-spring system shown, the mass, m, is displaced 0.09 m to the right of the equilibrium position and then released. What is most nearly the maximum velocity of m?

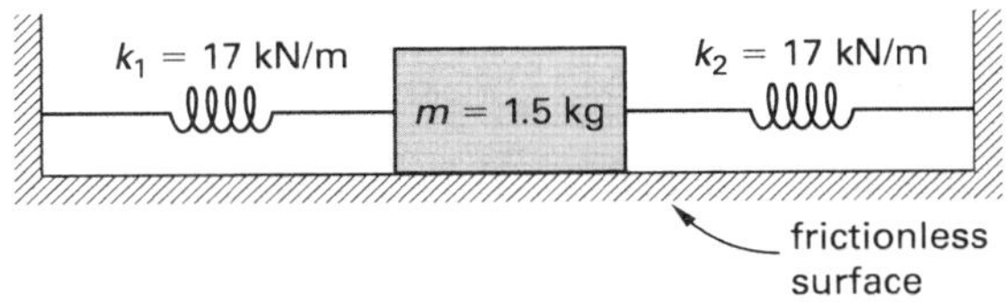

(A) 0.3 m/s

(B) 5 m/s

(C) 8 m/s

(D) 14 m/s

4. The 85 kg mass, m, shown is guided by a frictionless rail. The spring is compressed sufficiently and released, such that the mass barely reaches point B. The spring constant, k, is 1500 N/m.

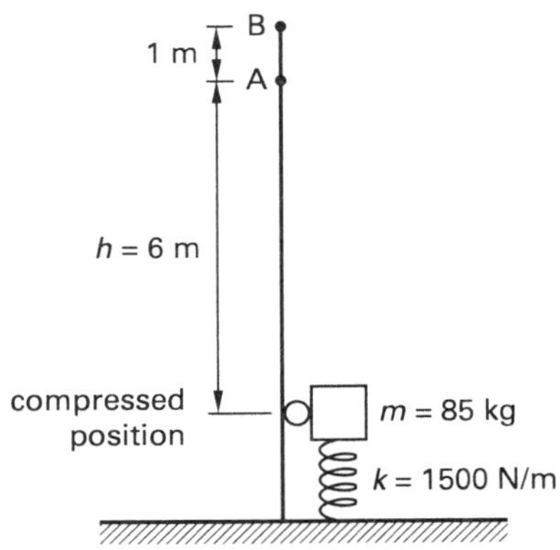

What is most nearly the velocity of the mass at point A?

(A) 3.1 m/s

(B) 4.4 m/s

(C) 9.8 m/s

(D) 20 m/s

5. The 170 kg mass, m, shown is guided by a frictionless rail. The spring is compressed sufficiently and released, such that the mass barely reaches point B.

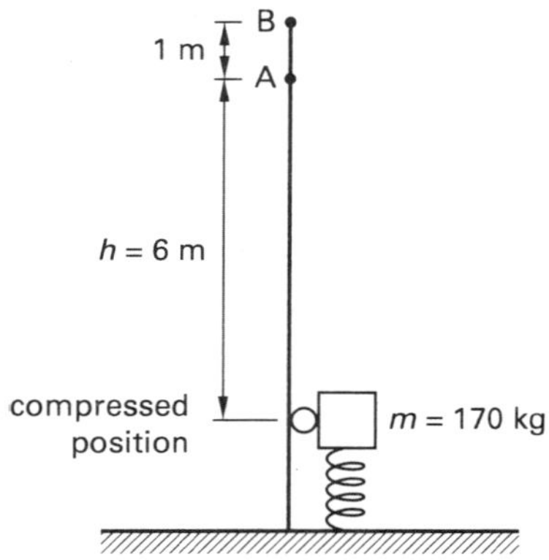

What is most nearly the kinetic energy of the mass at point A?

(A) 20 J

(B) 220 J

(C) 390 J

(D) 1700 J

6. The 40 kg mass, m, shown is guided by a frictionless rail. The spring constant, k, is 3000 N/m. The spring is compressed sufficiently and released, such that the mass barely reaches point A.

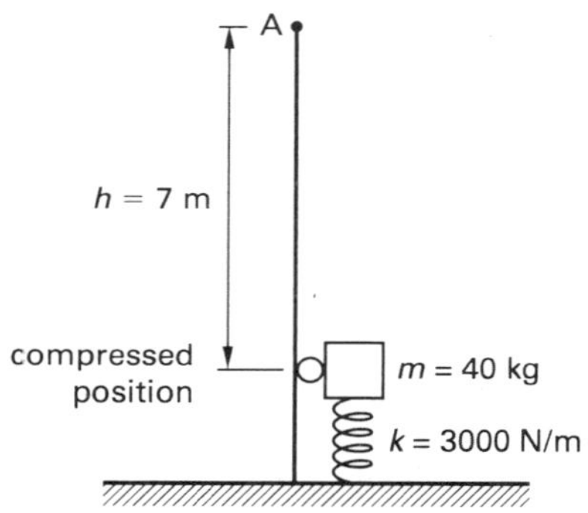

What is most nearly the initial spring compression?

(A) 0.96 m

(B) 1.3 m

(C) 1.4 m

(D) 1.8 m

7. A 1500 kg car traveling at 100 km/h is towing a 250 kg trailer. The coefficient of friction between the tires and the road is 0.8 for both the car and trailer. Approximately what energy is dissipated by the brakes if the car and trailer are braked to a complete stop?

(A) 96 kJ

(B) 390 kJ

(C) 580 kJ

(D) 680 kJ

8. A 12 kg aluminum box is dropped from rest onto a large wooden beam. The box travels 0.2 m before contacting the beam. After impact, the box bounces 0.05 m above the beams surface. Approximately what impulse does the beam impart on the box?

(A) 8.6 N·s

(B) 12 N·s

(C) 36 N·s

(D) 42 N·s

9. Two balls both have a mass of 8 kg and collide head on. The velocity of each ball at the time of collision is 18 m/s. The velocity of each ball decreases to 10 m/s in opposite directions after the collision. Approximately how much energy is lost in the collision?

(A) 0.57 kJ

(B) 0.91 kJ

(C) 1.8 kJ

(D) 2.3 kJ

10. The impulse-momentum principle is mostly useful for solving problems involving

(A) force, velocity, and time

(B) force, acceleration, and time

(C) velocity, acceleration, and time

(D) force, velocity, and acceleration

Engineering Sciences

11. A 50 kg cart rolls on a frictionless surface at 40 m/s. The cart is decelerated by a spring, initially at equilibrium, with a spring constant of 20 kN/m. Most nearly, what is the maximum deflection of the spring?

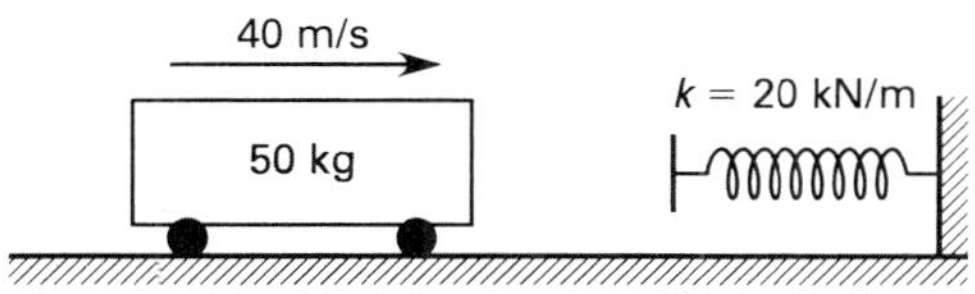

(A) 0.3 m

(B) 0.6 m

(C) 0.9 m

(D) 2.0 m

12. A pickup truck is traveling forward at 25 m/s. The bed is loaded with boxes whose coefficient of friction with the bed is 0.40. What is most nearly the shortest time that the truck can be brought to a stop such that the boxes do not shift?

(A) 2.3 s

(B) 4.7 s

(C) 5.9 s

(D) 6.4 s

SOLUTIONS

1. Due to the conservation of energy, the kinetic energy of the bullet before impact is equal to the potential energy of the spring-mass-bullet system at maximum compression.

$$E_{k,\text{bullet}} = E_{p,\text{system}}$$
$$\tfrac{1}{2}m_{\text{bullet}}\text{v}^2 = \tfrac{1}{2}kx^2$$

Rearrange this equation to solve for the velocity of the bullet.

$$\begin{aligned}\text{v} &= \sqrt{\frac{kx^2}{m_{\text{bullet}}}}\\ &= \sqrt{\frac{\left(53\ \frac{\text{kN}}{\text{cm}}\right)\left(1000\ \frac{\text{N}}{\text{kN}}\right)(3.4\ \text{cm})^2\left(1000\ \frac{\text{g}}{\text{kg}}\right)}{(100\ \text{g})\left(100\ \frac{\text{cm}}{\text{m}}\right)}}\\ &\quad\times\left(\frac{\left(60\ \frac{\text{s}}{\text{min}}\right)\left(60\ \frac{\text{min}}{\text{h}}\right)}{1000\ \frac{\text{m}}{\text{km}}}\right)\\ &= 891.1\ \text{km/h}\quad (890\ \text{km/h})\end{aligned}$$

The answer is (D).

2. The kinetic energy of the mass is

$$\begin{aligned}T &= m\text{v}^2/2\\ &= \frac{(0.05\ \text{kg})\left(0.4\ \frac{\text{m}}{\text{s}}\right)^2}{2}\\ &= 0.004\ \text{J}\end{aligned}$$

The potential energy of the mass is

$$\begin{aligned}U &= kx^2/2\\ &= \frac{\left(0.5\ \frac{\text{N}}{\text{m}}\right)(0.1\ \text{m})^2}{2}\\ &= 0.0025\ \text{J}\end{aligned}$$

The total energy of the mass is the sum of its kinetic and potential energies.

$$E = T + U = 0.004\ \text{J} + 0.0025\ \text{J} = 0.0065\ \text{J}$$

The answer is (C).

Engineering Sciences

3. At the moment before the mass is released (state 1), the kinetic energy of the mass is zero, and all the mass's energy is potential energy, U_1.

The maximum velocity will occur when the mass returns to the point of static equilibrium (state 2), where the deflection is zero, and the potential energy equals zero. At this point, all the mass's energy has been converted from potential energy to kinetic energy, T_2.

Set the pass's potential energy at the moment before release equal to the mass's kinetic energy at the point of equilibrium, and solve for the velocity. The displacement of each spring is

$$\begin{aligned} U_1 &= T_2 \\ \tfrac{1}{2}k_A x_A^2 + \tfrac{1}{2}k_B x_B^2 &= \tfrac{1}{2}m\text{v}^2 \\ \text{v} &= \sqrt{\frac{k_A x_A^2 + k_B x_B^2}{m}} \\ &= \sqrt{\frac{\left(\left(17\ \frac{\text{kN}}{\text{m}}\right)(0.09\ \text{m})^2 + \left(17\ \frac{\text{kN}}{\text{m}}\right)(0.09\ \text{m})^2\right)\left(1000\ \frac{\text{N}}{\text{kN}}\right)}{1.5\ \text{kg}}} \\ &= 13.5\ \text{m/s} \quad (14\ \text{m/s}) \end{aligned}$$

The answer is (D).

4. At point A, the energy of the mass is a combination of kinetic and gravitational potential energies. The total energy of the system is constant, and the kinetic energy at B is 0.

$$\begin{aligned} E_A &= E_B \\ U_A + T_A &= U_B \\ mgh + \frac{m\text{v}^2}{2} &= mg(h + 1\ \text{m}) \\ T_A &= mg(h + 1\ \text{m}) - mgh \\ &= mg(1\ \text{m}) \\ &= (85\ \text{kg})\left(9.81\ \frac{\text{m}}{\text{s}^2}\right)(1\ \text{m}) \\ &= 833.9\ \text{J} \end{aligned}$$

Therefore, the velocity of the mass at point A is

$$\begin{aligned} T_A &= \frac{m\text{v}^2}{2} = 833.9\ \text{J} \\ \text{v} &= \sqrt{\frac{2T_A}{m}} \\ &= \sqrt{\frac{(2)(833.9\ \text{J})}{85\ \text{kg}}} \\ &= 4.43\ \text{m/s} \quad (4.4\ \text{m/s}) \end{aligned}$$

The answer is (B).

5. At point A, the energy of the mass is a combination of kinetic and gravitational potential energies. The total energy of the system is constant, and the kinetic energy at B is 0.

$$\begin{aligned} E_A &= E_B \\ U_A + T_A &= U_B \\ mgh + \frac{m\text{v}^2}{2} &= mg(h + 1\ \text{m}) \\ T_A &= mg(h + 1\ \text{m}) - mgh \\ &= mg(1\ \text{m}) \\ &= (170\ \text{kg})\left(9.81\ \frac{\text{m}}{\text{s}^2}\right)(1\ \text{m}) \\ &= 1670\ \text{J} \quad (1700\ \text{J}) \end{aligned}$$

The answer is (D).

6. At the point just before the spring is released, all of the energy in the system is elastic potential energy; while at point A, all of the energy is potential energy due to gravity.

$$\begin{aligned} \frac{kx^2}{2} &= mgh \\ x &= \sqrt{\frac{2mgh}{k}} \\ &= \sqrt{\frac{(2)(40\ \text{kg})\left(9.81\ \frac{\text{m}}{\text{s}^2}\right)(7\ \text{m})}{3000\ \frac{\text{N}}{\text{m}}}} \\ &= 1.35\ \text{m} \quad (1.4\ \text{m}) \end{aligned}$$

The answer is (C).

Engineering Sciences

7. The original velocity of the car and trailer is

$$v = \frac{\left(100\ \frac{\text{km}}{\text{h}}\right)\left(1000\ \frac{\text{m}}{\text{km}}\right)}{\left(60\ \frac{\text{s}}{\text{min}}\right)\left(60\ \frac{\text{min}}{\text{h}}\right)} = 27.78\ \text{m/s}$$

Since the final velocity is zero, the energy dissipated is the original kinetic energy.

$$\begin{aligned} T &= \frac{mv^2}{2} = \frac{(1500\ \text{kg} + 250\ \text{kg})\left(27.78\ \frac{\text{m}}{\text{s}}\right)^2}{2} \\ &= 675\,154\ \text{J} \quad (680\ \text{kJ}) \end{aligned}$$

The answer is (D).

8. Initially, the box has potential energy only. (This takes the beams upper surface as the reference plane.) When the box reaches the beam, all of the potential energy will have been converted to kinetic energy.

$$\begin{aligned} mgh_1 &= \frac{mv_1^2}{2} \\ v_1 &= \sqrt{2gh_1} \\ &= \sqrt{(2)\left(9.81\ \frac{\text{m}}{\text{s}^2}\right)(0.2\ \text{m})} \\ &= 1.98\ \text{m/s} \quad [\text{downward}] \end{aligned}$$

When the box rebounds to its highest point, all of its remaining energy will be potential energy once again.

$$\begin{aligned} mgh_2 &= \frac{mv_2^2}{2} \\ v_2 &= \sqrt{2gh_2} \\ &= \sqrt{(2)\left(9.81\ \frac{\text{m}}{\text{s}^2}\right)(0.05\ \text{m})} \\ &= 0.99\ \text{m/s} \quad [\text{upward}] \end{aligned}$$

Use the impulse-momentum principle. (Downward is taken as the positive velocity direction.)

$$\begin{aligned} \text{Imp} &= \Delta p = m(v_1 - v_2) \\ &= (12\ \text{kg})\left(1.98\ \frac{\text{m}}{\text{s}} - \left(-0.99\ \frac{\text{m}}{\text{s}}\right)\right) \\ &= 35.66\ \text{N·s} \quad (36\ \text{N·s}) \end{aligned}$$

The answer is (C).

9. Each ball possesses kinetic energy before and after the collision. The velocity of each ball is reduced from |18 m/s| to |10 m/s|.

$$\begin{aligned} \Delta T &= T_2 - T_1 \\ &= (2)\left(\frac{m(v_2^2 - v_1^2)}{2}\right) \\ &= (2)\left(\frac{(8\ \text{kg})\left(\left(18\ \frac{\text{m}}{\text{s}}\right)^2 - \left(10\ \frac{\text{m}}{\text{s}}\right)^2\right)}{2}\right) \\ &= 1792\ \text{J} \quad (1.8\ \text{kJ}) \end{aligned}$$

The answer is (C).

10. Impulse is calculated from force and time. Momentum is calculated from mass and velocity. The impulse-momentum principle is useful in solving problems involving force, time, velocity, and mass.

The answer is (A).

11. Initially, with the spring at equilibrium (state 1), there is maximum kinetic energy in the cart and no potential energy in the spring. As the spring is compressed, the cart's kinetic energy is converted to potential energy. At the moment that the cart is stationary (state 2), the cart's kinetic energy is zero, having been entirely converted to potential energy in the spring.

Set the cart's kinetic energy at state 1, T_1, equal to the spring's potential energy at state 2, U_2, and solve for the maximum deflection of the spring, which occurs at state 2.

$$\begin{aligned} T_1 &= U_2 \\ \tfrac{1}{2}mv^2 &= \tfrac{1}{2}kx^2 \\ x &= \sqrt{\frac{mv^2}{k}} \\ &= \sqrt{\frac{(50\ \text{kg})\left(40\ \frac{\text{m}}{\text{s}}\right)^2}{\left(20\ \frac{\text{kN}}{\text{m}}\right)\left(1000\ \frac{\text{N}}{\text{kN}}\right)}} \\ &= 2.0\ \text{m} \end{aligned}$$

The answer is (D).

Engineering Sciences

12. The frictional force is the only force preventing the boxes from shifting. The forces on each box are its weight, the normal force, and the frictional force. The normal force on each box is equal to the box weight.

$$N = W = mg$$

The frictional force is

$$F_f = \mu N = \mu mg$$

Use the impulse-momentum principle. $v_2 = 0$. The frictional force is opposite of the direction of motion, so it is negative.

$$\begin{aligned} \text{Imp} &= \Delta p \\ F_f \Delta t &= m\Delta v \\ \Delta t &= \frac{m(v_2 - v_1)}{F_f} = \frac{-mv_1}{-\mu mg} = \frac{v_1}{\mu g} \\ &= \frac{25\ \frac{\text{m}}{\text{s}}}{(0.40)\left(9.81\ \frac{\text{m}}{\text{s}^2}\right)} \\ &= 6.37\ \text{s} \quad (6.4\ \text{s}) \end{aligned}$$

The answer is (D).

Engineering Sciences

19 Electrostatics

PRACTICE PROBLEMS

1. A system consists of two protons 3.4 μm apart in free space. Each proton has a charge of 1.6×10^{-19} C. A force moves the protons 200 nm closer together. Most nearly, what is the change in the potential energy of the system of two protons?

(A) 10^{-29} J

(B) 10^{-26} J

(C) 10^{-24} J

(D) 10^{-17} J

2. Particles A and B are 0.3 m apart in free space. Particle A has a charge of 1 C, and particle B has a charge of −8 C. Particles behave as point charges. Most nearly, what is the force on particle B due to particle A?

(A) 1×10^{13} N

(B) 8×10^{11} N

(C) 5×10^{10} N

(D) 9×10^{8} N

3. Two electrons are close to each other in free space. Both electrons experience a repulsive force of 1.0×10^{-15} N. Most nearly, how far apart are the electrons?

(A) 1.4×10^{-12} m

(B) 5.1×10^{-12} m

(C) 4.8×10^{-7} m

(D) 1.7×10^{-6} m

4. Two positive charges are separated in free space by a distance, x. The repulsive force between the charges is proportional to

(A) x

(B) x^2

(C) $\frac{1}{x^2}$

(D) $\frac{1}{x^3}$

5. An electron is moved against an electric field, which causes the electron's potential to increase by 10^6 V. Most nearly, what is the work performed in moving the electron?

(A) 4×10^{-19} J

(B) 5×10^{-16} J

(C) 8×10^{-14} J

(D) 2×10^{-13} J

6. A hollow metallic sphere surrounds a smaller metallic sphere of radius, r. A particle with a charge Q is located at the center of the smaller sphere. Of interest is the intensity of the electric field at a distance R from the surface of the smaller sphere. What function is proportional to this electric field intensity?

(A) $\frac{Q}{R^2}$

(B) $\frac{Q}{r^2}$

(C) $\frac{Q}{(R+r)^2}$

(D) $\frac{Q}{(R-r)^2}$

7. Three point charges, A, B, and C, are positioned in free space as shown.

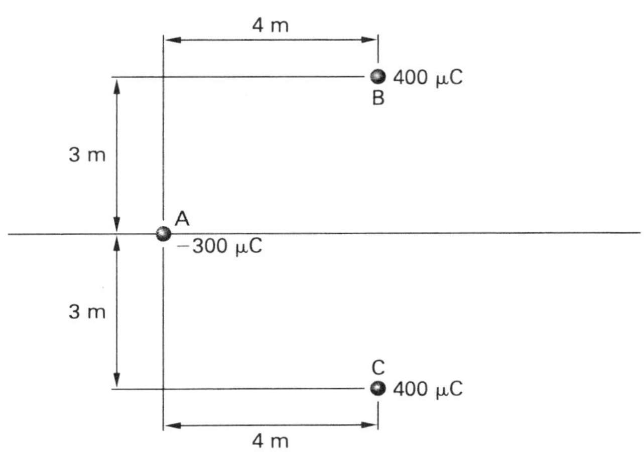

Most nearly, what are the magnitude and direction of the initial force that acts on charge A due to charges B and C?

(A) 35 N to the right

(B) 35 N to the left

(C) 69 N to the right

(D) 69 N to the left

8. Charge A is a point charge of –200 μC located in free space 10 cm from an infinite line charge of +100 μC/m. Charge B is a point charge of +4 μC located 5 cm from the line charge and directly between charge A and the line charge.

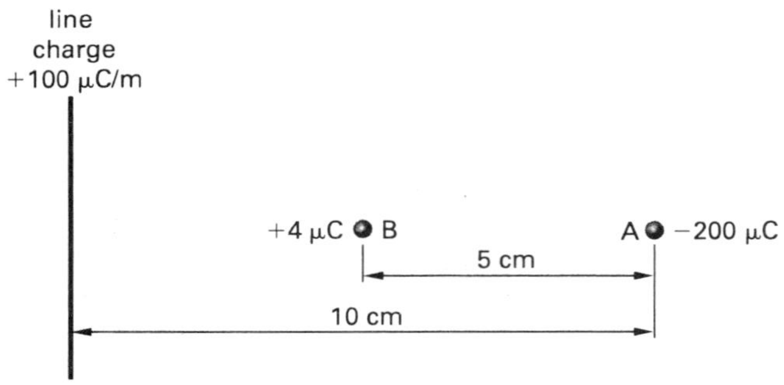

Most nearly, what is the attractive force on charge A due to the line charge and charge B?

(A) 1500 N

(B) 2200 N

(C) 3600 N

(D) 6500 N

9. An electric field is generated in an evacuated cube 10 cm on a side. The strength of the electric field is 0.5 V/m everywhere. Most nearly, what is the energy stored in the electric field?

(A) 1.1×10^{-15} J

(B) 2.2×10^{-15} J

(C) 5.5×10^{-13} J

(D) 1.5×10^{-12} J

SOLUTIONS

1. It doesn't make any difference whether both protons move or just one. The original separation distance is

$$r_1 = (3.4\ \mu\text{m})\left(10^{-6}\ \frac{\text{m}}{\mu\text{m}}\right) = 3.4 \times 10^{-6}\ \text{m}$$

The final separation distance is

$$\begin{aligned} r_2 &= r_1 - 200\ \text{nm} \\ &= 3.4 \times 10^{-6}\ \text{m} - \left(\frac{200\ \text{nm}}{10^9\ \frac{\text{nm}}{\text{m}}}\right) \\ &= 3.2 \times 10^{-6}\ \text{m} \end{aligned}$$

The change in potential energy is the same as the work needed to move the protons closer together. The work is

$$\begin{aligned} W &= -\int_{r_1}^{r_2} F dr = -\int_{r_1}^{r_2} \frac{Q_A Q_B}{4\pi\varepsilon r^2} dr \\ &= \left(\frac{Q_A Q_B}{4\pi\varepsilon}\right)\left(\frac{1}{r_2} - \frac{1}{r_1}\right) \\ &= \frac{(1.6 \times 10^{-19}\ \text{C})(1.6 \times 10^{-19}\ \text{C})}{4\pi\left(8.85 \times 10^{-12}\ \frac{\text{C}^2}{\text{N}\cdot\text{m}^2}\right)} \\ &\quad \times \left(\frac{1}{3.2 \times 10^{-6}\ \text{m}} - \frac{1}{3.4 \times 10^{-6}\ \text{m}}\right) \\ &= 4.23 \times 10^{-24}\ \text{J} \quad (10^{-24}\ \text{J}) \end{aligned}$$

The answer is (C).

2. The force on particle B due to particle A is the same as the force on particle A due to particle B. From Coulomb's law, the force between the particles is

$$\begin{aligned} F &= \frac{Q_1 Q_2}{4\pi\varepsilon_o r^2} = \frac{(1\ \text{C})(-8\ \text{C})}{4\pi\left(8.85 \times 10^{-12}\frac{\text{C}^2}{\text{N}\cdot\text{m}^2}\right)(0.3\ \text{m})^2} \\ &= -7.99 \times 10^{11}\ \text{N} \quad (8 \times 10^{11}\ \text{N attractive}) \end{aligned}$$

The answer is (B).

3. Each electron has a charge of -1.6×10^{-19} C, so $Q_1 = Q_2$. Use Coulomb's law, and solve for the distance between the two electrons.

$\varepsilon_o = 8.85 \times 10^{-12}\ \text{C}^2/\text{N·m}^2$. Also, for an electron, $Q = 1.6 \times 10^{-19}$ C. Solving for r,

$$\begin{aligned} F &= \frac{Q_1 Q_2}{4\pi\varepsilon_o r^2} \\ r &= Q\sqrt{\frac{1}{4\pi\varepsilon_o F}} \\ &= (1.6 \times 10^{-19}\ \text{C})\sqrt{\frac{1}{4\pi\left(8.85 \times 10^{-12}\ \frac{\text{C}^2}{\text{N·m}^2}\right)(1 \times 10^{-15}\ \text{N})}} \\ &= 4.78 \times 10^{-7}\ \text{m} \quad (4.8 \times 10^{-7}\ \text{m}) \end{aligned}$$

The answer is (C).

4. The force must overcome the repulsive force between the charges. The repulsive force is given by Coulomb's law.

$$F = \frac{Q_1 Q_2}{4\pi\varepsilon_o r^2}$$

F is proportional to the inverse of r^2, or $1/x^2$.

The answer is (C).

5. The work done in moving an object with charge Q parallel to an electric field so as to cause a change in potential of ΔV is

$$\begin{aligned} W &= -Q\Delta V \\ &= -(-1.6 \times 10^{-19}\ \text{C})(10^6\ \text{V}) \\ &= 1.6 \times 10^{-13}\ \text{J} \qquad (2 \times 10^{-13}\ \text{J}) \end{aligned}$$

The answer is (D).

6. The central charge sets up an electric field within the smaller sphere, which in turn induces an electric field within the larger sphere. The electric field intensity is

$$E = \frac{F}{Q_2} = \frac{Q_1 Q_2}{4\pi\varepsilon(R+r)^2 Q_2} \propto \frac{Q_1}{(R+r)^2}$$

The electric field intensity is proportional to the square of the distance from the smaller sphere's center.

The answer is (C).

7. Opposite charges attract, so the direction of the impending motion (and, equivalently, of the force) is toward charges B and C. Charges B and C are equal in magnitude but in opposite vertical directions from charge A, so the two charges combined will cause zero net vertical force on charge A. The total horizontal force on charge A created by the two other charges will be to the right.

From the Pythagorean theorem, the distance between charge A and either of the other charges is

$$\begin{aligned} c &= \sqrt{a^2 + b^2} \\ &= \sqrt{(4\ \text{m})^2 + (3\ \text{m})^2} \\ &= 5\ \text{m} \end{aligned}$$

From Coulomb's law, the force acting on charge A due to either of the other two charges is

$$\begin{aligned} F &= \frac{Q_1 Q_2}{4\pi\varepsilon r^2} \\ &= \frac{(-300\ \mu\text{C})(400\ \mu\text{C})}{4\pi\left(8.85 \times 10^{-12}\ \frac{\text{C}^2}{\text{N·m}^2}\right)(5\ \text{m})^2\left(10^6\ \frac{\mu C}{C}\right)^2} \\ &= -43.16\ \text{N} \end{aligned}$$

The horizontal component of the distance to each charge is 4/5 of the actual distance to each charge, so the horizontal component of the force due to each charge is 4/5 of the force in the direction of the charge.

$$F_{\text{horiz}} = \tfrac{4}{5}F = \left(\frac{4}{5}\right)(-43.16\ \text{N}) = -34.53\ \text{N}$$

The total horizontal force acting on charge A due to charges B and C is twice that due to either charge alone, so

$$\begin{aligned} F_{\text{total}} &= 2F_{\text{horiz}} = (2)(-34.53\ \text{N}) \\ &= -69.06\ \text{N} \quad (69\ \text{N to the right}) \end{aligned}$$

The answer is (C).

8. The total force on charge A is the sum of the forces due to the line charge and charge B.

The strength of the electric field created by the line charge is

$$\begin{aligned} E_L &= \frac{\rho_L}{2\pi\varepsilon r} = \frac{\dfrac{100\ \frac{\mu\text{C}}{\text{m}}}{10^6\ \frac{\mu\text{C}}{\text{C}}}}{2\pi\left(8.85 \times 10^{-12}\ \frac{\text{C}^2}{\text{N·m}^2}\right)\left(\dfrac{10\ \text{cm}}{100\ \frac{\text{cm}}{\text{m}}}\right)} \\ &= 1.80 \times 10^7\ \text{N/C} \end{aligned}$$

The force on charge A in the electric field created by the line charge is

$$F_L = QE = \left(\frac{-200\ \mu\text{C}}{10^6\ \frac{\mu\text{C}}{\text{C}}}\right)\left(1.80\times 10^7\ \frac{\text{N}}{\text{C}}\right) = -3600\ \text{N}$$

The force on charge A in the electric field created by charge B is

$$F_\text{B} = \frac{Q_\text{B}Q_\text{A}}{4\pi\varepsilon r^2} = \frac{\left(\frac{4\ \mu\text{C}}{10^6\ \frac{\mu\text{C}}{\text{C}}}\right)\left(\frac{-200\ \mu\text{C}}{10^6\ \frac{\mu\text{C}}{\text{C}}}\right)}{4\pi\left(8.85\times 10^{-12}\ \frac{\text{C}^2}{\text{N}\cdot\text{m}^2}\right)\left(\frac{5\ \text{cm}}{100\ \frac{\text{cm}}{\text{m}}}\right)^2} = -2877\ \text{N}$$

The total force on charge A from the line charge and charge B is

$$F_\text{total} = F_L + F_\text{B} = -3600\ \text{N} + (-2877\ \text{N}) = -6477\ \text{N} \quad (6500\ \text{N attractive})$$

The answer is (D).

Electromagnetics

9. The vacuum within the cube has a permittivity of 8.85×10^{-12} F/m. The energy stored in the electric field is

$$W_E = (1/2)\iiint_V \varepsilon|\mathbf{E}|^2\,dV = \frac{\varepsilon E}{2}\iiint_V dV = \left(\frac{\left(8.85\times 10^{-12}\ \frac{\text{F}}{\text{m}}\right)\left(0.5\ \frac{\text{V}}{\text{m}}\right)^2}{2}\right)\left(\frac{10\ \text{cm}}{100\ \frac{\text{cm}}{\text{m}}}\right)^3 = 1.11\times 10^{-15}\ \text{J} \quad (1.1\times 10^{-15}\ \text{J})$$

The answer is (A).

20 Magnetism and Magnetostatics

PRACTICE PROBLEMS

1. An electromagnet has an iron core with a permeability of 6×10^{-5} T·m/A. The magnetic field through the center of the electromagnet is nearly uniform and has a strength of 7.96×10^{-3} A/m. When the iron core is removed, the air inside the electromagnet has a permeability of $4\pi \times 10^{-7}$ T·m/A. Most nearly, what is the magnetic flux density of the field through the iron core?

(A) 1.0×10^{-8} T

(B) 4.8×10^{-7} T

(C) 7.8×10^{-7} T

(D) 15×10^{-6} T

2. A uniform magnetic field with a magnetic flux density of 5.5×10^{-4} T passes through an evacuated cube with sides measuring 0.125 m, as shown. What is most nearly the magnetic energy contained in the cube?

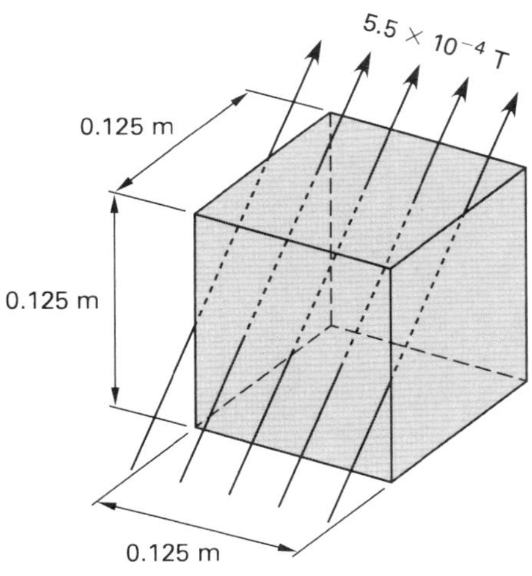

(A) 1.1×10^{-6} J

(B) 8.6×10^{-6} J

(C) 2.4×10^{-4} J

(D) 4.7×10^{-4} J

3. Two wires are oriented in free space as shown. Wire A is parallel to the z-axis and carries 2 mA of current flowing in the positive z-direction. Wire B is parallel to the y-axis and carries 3 mA of current flowing in the positive y-direction. The wires are 10 cm apart at their closest point.

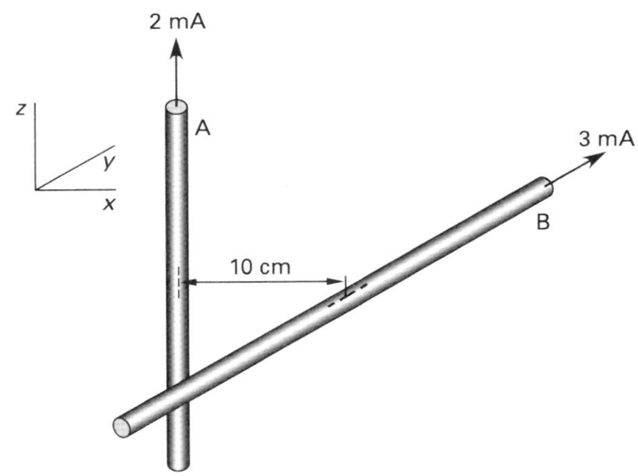

Most nearly, what is the magnetic field strength halfway between the wires at the point where they are closest?

(A) $(2.0 \times 10^{-2}\ \text{A/m})\mathbf{j} + (3.0 \times 10^{-2}\ \text{A/m})\mathbf{k}$

(B) $(3.2 \times 10^{-3}\ \text{A/m})\mathbf{i} + (4.8 \times 10^{-3}\ \text{A/m})\mathbf{j}$

(C) $(6.4 \times 10^{-3}\ \text{A/m})\mathbf{j} + (9.6 \times 10^{-3}\ \text{A/m})\mathbf{k}$

(D) $(9.6 \times 10^{-3}\ \text{A/m})\mathbf{j} + (6.4 \times 10^{-3}\ \text{A/m})\mathbf{k}$

4. A 0.3 m long wire runs parallel to the z-axis as shown. The wire carries a 4 mA current flowing in the positive z-direction. When the wire is exposed to a uniform magnetic field parallel to the x-y plane, the wire experiences a force of

$$(3\ \text{N})\left(\frac{\sqrt{2}}{2}\right)\mathbf{i} + (3\ \text{N})\left(\frac{\sqrt{2}}{2}\right)\mathbf{j}$$

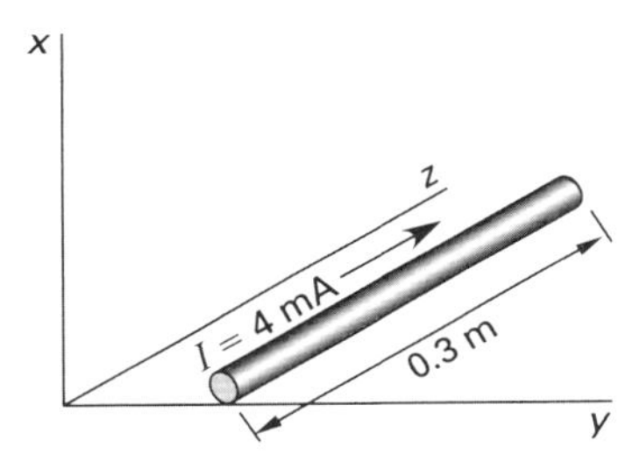

Most nearly, what is the magnetic flux density of the uniform magnetic field?

(A) 530 T

(B) 1800 T

(C) 2500 T

(D) 3000 T

5. A 1.27 m length of wire carrying a current of 5 A is exposed to a uniform magnetic field with a magnetic flux density of 0.17 T. The wire is straight, and the direction of the current is perpendicular to the direction of the magnetic field. Most nearly, what is the force that the magnetic field exerts on the wire?

(A) 0.2 N

(B) 0.5 N

(C) 0.8 N

(D) 1.1 N

6. Two parallel conductors in free space are 5 cm apart as shown. Each conductor carries 50 A in the same direction.

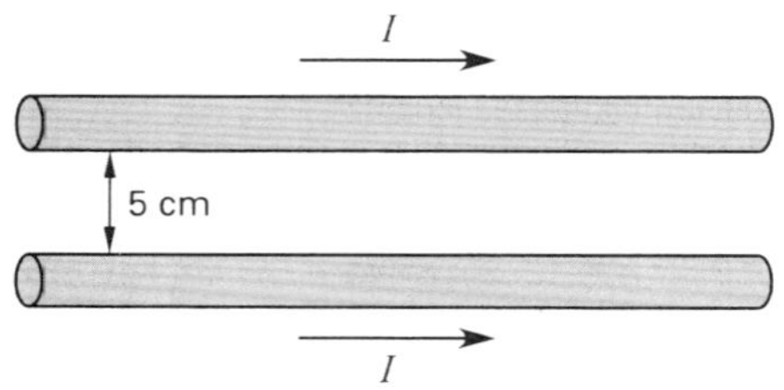

Most nearly, what is the magnetic force on a 1 m length of one conductor?

(A) 0.01 N

(B) 0.02 N

(C) 0.03 N

(D) 0.04 N

7. Two parallel conductors are located 1 m apart in free space. Each conductor carries a current of 10 A, but the two currents flow in opposite directions. Most nearly, what is the force per meter exerted by one conductor on the other?

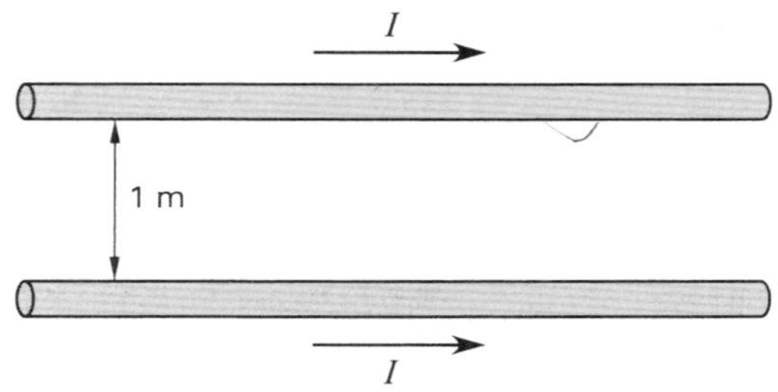

(A) 4.0×10^{-7} N/m

(B) 2.0×10^{-7} N/m

(C) 1.0×10^{-6} N/m

(D) 2.0×10^{-5} N/m

8. A magnetic field has a constant strength of 0.5 A/m within an evacuated cube measuring 10 cm per side. Most nearly, what is the magnetic energy contained within the cube?

(A) 1.6×10^{-10} J

(B) 5.5×10^{-9} J

(C) 3.5×10^{-9} J

(D) 1.1×10^{-9} J

SOLUTIONS

1. The permeability of the electromagnet when the core is removed is irrelevant. Use the equation for magnetic field strength, and solve for the magnetic flux density.

$$\begin{aligned}\mathbf{H} &= \frac{\mathbf{B}}{\mu}\\ \mathbf{B} &= \mathbf{H}\mu\\ &= \left(7.96\times 10^{-3}\ \frac{\text{A}}{\text{m}}\right)\left(6\times 10^{-5}\ \frac{\text{T·m}}{\text{A}}\right)\\ &= 4.77\times 10^{-7}\ \text{T}\quad (4.8\times 10^{-7}\ \text{T})\end{aligned}$$

The answer is (B).

2. The magnetic field strength is

$$\begin{aligned}\mathbf{H} &= \frac{\mathbf{B}}{\mu}\\ &= \frac{5.5\times 10^{-4}\ \text{T}}{4\pi\times 10^{-7}\ \frac{\text{T·m}}{\text{A}}}\\ &= 437.7\ \text{A/m}\end{aligned}$$

Find the energy stored in the magnetic field. The magnetic field strength is constant throughout the cube.

$$\begin{aligned}W_H &= \tfrac{1}{2}\iiint_V \mu\,|\mathbf{H}|^2\,dv\\ &= \tfrac{1}{2}\mu\,|\mathbf{H}|^2 \iiint_V dv\\ &= \left(\frac{1}{2}\right)\left(4\pi\times 10^{-7}\ \frac{\text{T·m}}{\text{A}}\right)\left(437.7\ \frac{\text{A}}{\text{m}}\right)^2(0.125\ \text{m})^3\\ &= 2.35\times 10^{-4}\ \text{J}\quad (2.4\times 10^{-4}\ \text{J})\end{aligned}$$

The answer is (C).

3. From the right-hand rule, wire A's magnetic field is in the positive y-direction, so $\mathbf{a}_{\phi\text{A}} = \mathbf{j}$. Wire B's magnetic field is in the positive z-direction, so, $\mathbf{a}_{\phi\text{B}} = \mathbf{k}$. At a point halfway between the wires where they are closest, the radial distance from each wire is $10\ \text{cm}/2 = 5\ \text{cm}$.

The magnetic field strength of wire A is

$$\begin{aligned}\mathbf{H}_\text{A} &= \frac{I\mathbf{a}_{\phi\text{A}}}{2\pi r}\\ &= \left(\frac{I}{2\pi r}\right)\mathbf{a}_{\phi\text{A}}\\ &= \left(\frac{\dfrac{2\ \text{mA}}{1000\ \dfrac{\text{mA}}{\text{A}}}}{2\pi\left(\dfrac{5\ \text{cm}}{100\ \dfrac{\text{cm}}{\text{m}}}\right)}\right)\mathbf{j}\\ &= (6.37\times 10^{-3}\ \text{A/m})\mathbf{j}\quad ((6.4\times 10^{-3}\ \text{A/m})\mathbf{j})\end{aligned}$$

The magnetic field strength of wire B is

$$\begin{aligned}\mathbf{H}_\text{B} &= \frac{I\mathbf{a}_{\phi\text{B}}}{2\pi r}\\ &= \left(\frac{I}{2\pi r}\right)\mathbf{a}_{\phi\text{B}}\\ &= \left(\frac{\dfrac{3\ \text{mA}}{1000\ \dfrac{\text{mA}}{\text{A}}}}{2\pi\left(\dfrac{5\ \text{cm}}{100\ \dfrac{\text{cm}}{\text{m}}}\right)}\right)\mathbf{k}\\ &= (9.55\times 10^{-3}\ \text{A/m})\mathbf{k}\quad ((9.6\times 10^{-3}\ \text{A/m})\mathbf{k})\end{aligned}$$

The total magnetic field strength is

$$\begin{aligned}\mathbf{H} &= \mathbf{H}_\text{A} + \mathbf{H}_\text{B}\\ &= (6.4\times 10^{-3}\ \text{A/m})\mathbf{j} + (9.6\times 10^{-3}\ \text{A/m})\mathbf{k}\end{aligned}$$

The answer is (C).

4. The magnitude of the force experienced by the wire is

$$\begin{aligned}F &= \sqrt{F_x^2 + F_y^2 + F_z^2}\\ &= \sqrt{\left((3\ \text{N})\left(\frac{\sqrt{2}}{2}\right)\right)^2 + \left((3\ \text{N})\left(\frac{\sqrt{2}}{2}\right)\right)^2 + 0\ \text{N}}\\ &= 3\ \text{N}\end{aligned}$$

Electromagnetics

Use the equation for force in a uniform magnetic field, and solve for the magnetic flux density.

$$\mathbf{F} = I\mathbf{L} \times \mathbf{B}$$
$$\mathbf{B} = \frac{\mathbf{F}}{I\mathbf{L}} = \frac{3\ \text{N}}{(4.0 \times 10^{-3}\ \text{A})(0.3\ \text{m})} = 2500\ \text{T}$$

The answer is (C).

5. The force that the magnetic field exerts on the wire is

$$\mathbf{F} = I\mathbf{L} \times \mathbf{B} = (5\ \text{A})(1.27\ \text{m})(0.17\ \text{T}) = 1.08\ \text{N} \quad (1.1\ \text{N})$$

The answer is (D).

6. Use the equation for magnetic field strength on a wire, and solve for the magnetic flux density.

$$\mathbf{H} = \frac{\mathbf{B}}{\mu} = \frac{I}{2\pi r}$$
$$\mathbf{B} = \frac{\mu I}{2\pi r} = \frac{\left(4\pi \times 10^{-7}\ \frac{\text{T}\cdot\text{m}}{\text{A}}\right)(50\ \text{A})}{2\pi\left(\frac{5\ \text{cm}}{100\ \frac{\text{cm}}{\text{m}}}\right)} = 2.0 \times 10^{-4}\ \text{T}$$

The force of the magnetic field is

$$\mathbf{F} = I\mathbf{L} \times \mathbf{B} = (50\ \text{A})(1\ \text{m})(2.0 \times 10^{-4}\ \text{T}) = 0.01\ \text{N}$$

The answer is (A).

7. Each conductor produces a magnetic field at the other conductor. Use the equation for magnetic field strength on a wire, and solve for the magnetic flux density.

$$\mathbf{H} = \frac{\mathbf{B}}{\mu} = \frac{I}{2\pi r}$$
$$\mathbf{B} = \frac{\mu I}{2\pi r} = \frac{\left(4\pi \times 10^{-7}\ \frac{\text{H}}{\text{m}}\right)(10\ \text{A})}{2\pi(1\ \text{m})} = 2.0 \times 10^{-6}\ \text{Wb/m}^2$$

Use the equation for force in a uniform magnetic field, and solve for the force per unit length.

$$\mathbf{F} = I\mathbf{L} \times \mathbf{B}$$
$$\frac{\mathbf{F}}{\mathbf{L}} = I \times \mathbf{B} = (10\ \text{A})\left(2.0 \times 10^{-6}\ \frac{\text{Wb}}{\text{m}^2}\right) = 2.0 \times 10^{-5}\ \text{N/m}$$

The answer is (D).

8. Use the equation for the energy stored in a uniform magnetic field. The magnetic energy contained within the cube is

$$U = \frac{\mu H^2}{2} V = \left(\frac{\left(4\pi \times 10^{-7}\ \frac{\text{H}}{\text{m}}\right)\left(0.5\ \frac{\text{A}}{\text{m}}\right)^2}{2}\right)\left(\frac{10\ \text{cm}}{100\ \frac{\text{cm}}{\text{m}}}\right)^3 = 1.571 \times 10^{-10}\ \text{J} \quad (1.6 \times 10^{-10}\ \text{J})$$

The answer is (A).

Electromagnetics

21 Maxwell's Equations and Related Laws

PRACTICE PROBLEMS

1. A straight conducting wire with a diameter of 1 mm runs along the z-axis. The magnetic field strength outside the wire is $(0.02/\rho)\mathbf{a}_\phi$ A/m. ρ is the distance from the center of the wire. Of interest is the total magnetic flux within an area from $\rho = 0.5$ mm to 2 cm and $z = 0$ to 4 m. Most nearly, that magnetic flux is

(A) 9.3×10^{-8} Wb

(B) 1.4×10^{-7} Wb

(C) 3.7×10^{-7} Wb

(D) 3.0×10^{-4} Wb

2. In the differential form of Faraday's law of induction, $\nabla \times \mathbf{E} = -\partial\mathbf{B}/\partial t$, which units are associated with the magnetic field term, $-\partial\mathbf{B}/\partial t$?

(A) amperes

(B) volts

(C) volts per meter

(D) volts per square meter

3. In Ampere's law, what is represented by the term $\partial\mathbf{D}/\partial t$?

(A) current density

(B) displacement current

(C) electric flux density

(D) magnetic field strength

4. In the differential form of Maxwell's equation, $\nabla \cdot x = 0$, what does x represent?

(A) **B**, magnetic flux density

(B) **D**, electric field density

(C) **E**, electric field strength

(D) **J**, current density

5. The electric field 1 m from a point charge in free space has a strength of 1.8×10^5 V/m. Most nearly, the charge is

(A) 20 μC

(B) 60 μC

(C) 80 μC

(D) 100 μC

SOLUTIONS

1. Use the equation for magnetic field strength to find the magnetic flux density as a function of distance.

$$\begin{aligned}\mathbf{H} &= \frac{\mathbf{B}}{\mu}\\ \mathbf{B} &= \mu\mathbf{H}\\ &= \left(4\pi\times 10^{-7}\ \frac{\text{H}}{\text{m}}\right)\left(\frac{0.02\ \text{A}}{\rho}\right)\mathbf{a}_\phi\\ &= \left(8\pi\times 10^{-9}\ \frac{\text{Wb}}{\text{m}}\right)\left(\frac{1}{\rho}\right)\mathbf{a}_\phi\end{aligned}$$

Use Gauss' law for magnetism to find the magnetic flux. Integrate the flux density relation over the area of interest.

$$\oint_{S_V}\mathbf{B}\cdot d\mathbf{S} = 0$$

$$\begin{aligned}\phi &= \iint \mathbf{B}\cdot d\mathbf{S}\\ &= \iint\left(8\pi\times 10^{-9}\ \frac{\text{Wb}}{\text{m}}\left(\frac{1}{\rho}\right)\mathbf{a}_\phi\right)(dz\ d\rho\mathbf{a}_\phi)\\ &= 8\pi\times 10^{-9}\ \frac{\text{Wb}}{\text{m}}\int_{0\,\text{m}}^{4\,\text{m}} dz\int_{0.0005\,\text{m}}^{0.02\,\text{m}}\frac{d\rho}{\rho}\\ &= \left(8\pi\times 10^{-9}\ \frac{\text{Wb}}{\text{m}}\right)(4\ \text{m} - 0\ \text{m})\ln\frac{0.02\ \text{m}}{0.0005\ \text{m}}\\ &= 3.7\times 10^{-7}\ \text{Wb}\end{aligned}$$

The answer is (C).

2. The units for magnetic flux density are webers per square meter. Taking the partial derivative with time gives units of Wb/m^2·s. Since that is not one of the answers, simplify the units. Webers are equal to volt-seconds, the result being V·s/m^2·s = V/m^2 (volts per square meter).

The answer is (D).

3. In Ampere's law, the term $\partial\mathbf{D}/\partial t$ represents the displacement current.

The answer is (B).

4. In Maxwell's differential, or point form, equations, the divergence operator is used with the electric and magnetic fields. In Gauss' law for electricity, $\nabla\cdot\mathbf{D} = \rho$, the divergence of the electric flux density, $\mathbf{D}$, equals the charge enclosed, ρ. In Gauss' law for magnetism, $\nabla\cdot\mathbf{B} = 0$, the divergence of the magnetic flux density, $\mathbf{B}$, is zero, indicating that there are no monopoles, or magnetic charges. x represents $\mathbf{B}$.

The answer is (A).

5. The charge can be determined by integrating a spherical surface centered at the point charge.

$$\begin{aligned}Q_{\text{encl}} &= \oint_S \varepsilon\mathbf{E}\cdot d\mathbf{S}\\ &= \varepsilon_0\mathbf{E}\int_0^\pi r^2\sin\theta\ d\theta\int_0^\pi d\phi\\ &= 2\pi\varepsilon_0\mathbf{E}r^2\int_0^\pi \sin\theta\ d\theta\\ &= 2\pi\varepsilon_0\mathbf{E}r^2 - \cos\theta\big|_0^\pi\\ &= 2\pi\varepsilon_0\mathbf{E}r^2(-(-1)-(-1))\\ &= 4\pi\varepsilon_0\mathbf{E}r^2\\ &= 4\pi\left(8.854\times 10^{-12}\ \frac{\text{C}^2}{\text{N·m}^2}\right)\left(1.8\times 10^5\ \frac{\text{V}}{\text{m}}\right)(1\ \text{m})^2\\ &= 20\times 10^{-6}\ \text{C}\quad (20\ \mu\text{C})\end{aligned}$$

(Integration can be avoided by remembering that the area of a sphere is $4\pi r^2$.)

Alternative Solution

Use the equation for electric field intensity to find the point charge.

$$\begin{aligned}\mathbf{E} &= \frac{Q_1}{4\pi\varepsilon r^2}\mathbf{a}_{r12}\\ Q_1 &= 4\pi\varepsilon_0 r^2\\ &= 4\pi\left(8.854\times 10^{-12}\ \frac{\text{C}^2}{\text{N·m}^2}\right)\left(1.8\times 10\ \frac{\text{V}}{\text{m}}\right)(1\ \text{m})^2\\ &= 20\times 10^{-6}\ \text{C}\quad (20\ \mu\text{C})\end{aligned}$$

The answer is (A).

22 Electromagnetic Wave Propagation and Compatibility

PRACTICE PROBLEMS

1. Most nearly, the equivalent decibel gain for a voltage gain of 40,000 is

(A) 40 dB

(B) 46 dB

(C) 92 dB

(D) 100 dB

2. An underwater transducer detects a 200 Hz pressure wave signal. The speed of sound in water is 2.24×10^8 m/s. Most nearly, the wavelength of the signal is

(A) 10 mi

(B) 110 mi

(C) 700 mi

(D) 3100 mi

3. When power is doubled, the value of the increase in decibels is

(A) 1 dB

(B) 2 dB

(C) 3 dB

(D) 6 dB

4. A signal gain meter uses 1 mV as its standard reference signal strength value. If the measured gain is $-20\ \text{dB}_{\text{mV}}$, the signal strength is

(A) 0.01 mV

(B) 0.1 mV

(C) 10 mV

(D) 100 mV

5. Which of the following is NOT a form of electromagnetic interference?

(A) catalytic coupling

(B) radiated emissions coupling

(C) magnetic coupling

(D) conductive coupling

6. Which of the following can be used to improve overall electromagnetic susceptibility hardening against radiated emissions?

(A) increased switching speed

(B) filtering circuits

(C) shielded housings

(D) fuses and circuit breakers

Electromagnetics

SOLUTIONS

1. Voltage gain is a ratio of two voltages. The gain in decibels that is equivalent to a voltage gain of 40,000 is

$$\begin{aligned} G &= 20\log_{10}\frac{V_{\text{std}}}{V_{\text{test}}} \\ &= 20\log_{10}40{,}000 \\ &= 92.04\ \text{dB} \quad (92\ \text{dB}) \end{aligned}$$

The answer is (C).

2. The wavelength is

$$\begin{aligned} \lambda &= \frac{U}{f} \\ &= \frac{2.24\times 10^8\ \frac{\text{m}}{\text{s}}}{(200\ \text{Hz})\left(1609.3\ \frac{\text{m}}{\text{mi}}\right)} \\ &= 696\ \text{mi} \quad (700\ \text{mi}) \end{aligned}$$

The answer is (C).

3. The doubling of power is a common frame of reference in engineering. Regardless of the reference power, when power, P, is doubled, the value of the increase in decibels is

$$\begin{aligned} X_{\text{dB}} &= 10\log_{10}\left(\frac{2P}{P}\right) = 10\log_{10}2 \\ &= 3\ \text{dB} \end{aligned}$$

The answer is (C).

4. Rearrange the equation for gain to solve for the signal strength.

$$\begin{aligned} G &= 20\log_{10}\frac{V_{\text{signal}}}{V_{\text{ref}}} \\ V_{\text{signal}} &= V_{\text{ref}}10^{G/20} \\ &= (1\ \text{mV})(10^{-20\ \text{dB}_{\text{mV}}/20\ \text{dB}_{\text{mV}}}) \\ &= 0.1\ \text{mV} \end{aligned}$$

The answer is (B).

5. Radiated emissions, magnetic effects, and conductive coupling from a source can all interfere with the normal operation of a victim system, so these are forms of electromagnetic interference. Catalytic coupling is a chemical process in which a metal catalyst aids in a chemical reaction between two hydrocarbons; it is not a form of electromagnetic interference.

The answer is (A).

6. Shielded housings are conductive enclosures around circuits that are grounded; they divert emissions to ground, so they improve electromagnetic compatibility.

High switching speeds can cause some electromagnetic interference, so increasing the switching speed will not improve electromagnetic susceptibility hardening. Filtering circuits are connected in the conductive path; they reduce electromagnetic interference that is outside the frequency range of the desired signal, so they do not improve electromagnetic susceptibility hardening. Fuses and circuit breakers protect against ground faults; they do not improve electromagnetic susceptibility hardening.

The answer is (C).

Electromagnetics

23 Direct-Current Circuits

PRACTICE PROBLEMS

1. An ideal battery, an ideal current source, and three resistors are assembled into a circuit as shown.

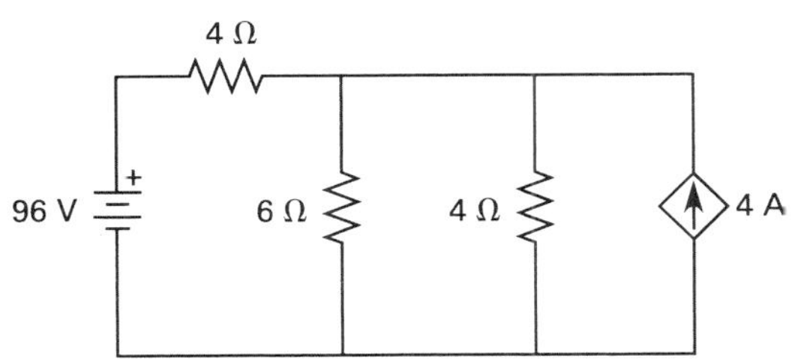

Most nearly, the current flowing through the 6 Ω resistor is

(A) 7.0 A

(B) 7.5 A

(C) 10 A

(D) 12 A

2. A circuit with two terminals is shown.

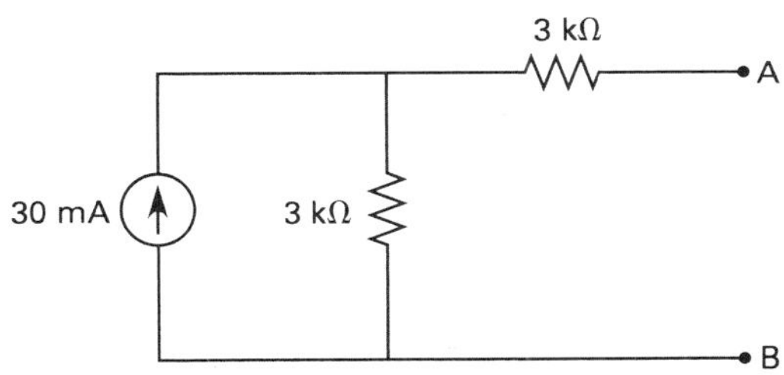

The Thevenin equivalent for this circuit is

(A) 30 V in series with 1.5 kΩ

(B) 45 V in series with 1.5 kΩ

(C) 90 V in series with 3 kΩ

(D) 90 V in series with 6 kΩ

3. The circuit shown includes a voltage source, a current source, and two resistors.

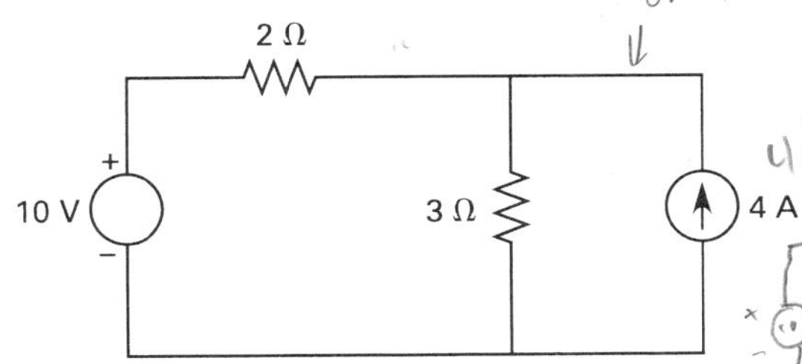

Most nearly, the current through the 2 Ω resistor is

(A) 0.4 A, right to left

(B) 0.4 A, left to right

(C) 1.7 A, right to left

(D) 1.7 A, left to right

4. A meter movement has an internal resistance of 5 kΩ and a full-scale current of 50 μA. The movement is to be used within a voltmeter with a full-scale voltage of 20 V. A resistance in series with the movement will be used to limit full-scale meter indications. Most nearly, the value of the series resistance is

(A) 2 kΩ

(B) 5 kΩ

(C) 400 kΩ

(D) 500 kΩ

5. A voltmeter is used to measure the voltage across a resistor. The voltmeter is connected in parallel with the resistor. To obtain an accurate voltage reading, the internal resistance of the voltmeter must be

(A) much smaller than the resistance whose voltage is being measured

(B) approximately equal to the resistance whose voltage is being measured

(C) approximately equal to 1.7 times the resistance whose voltage is being measured

(D) much greater than the resistance whose voltage is being measured

6. A circuit with two terminals is shown.

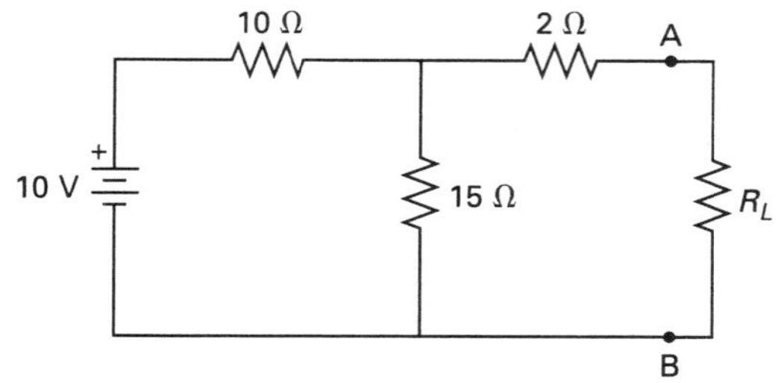

Most nearly, the resistance at R_L that will allow maximum power transfer through terminals A and B is

(A) 2.0 Ω
(B) 8.0 Ω
(C) 15 Ω
(D) 17 Ω

7. A circuit with two batteries is shown.

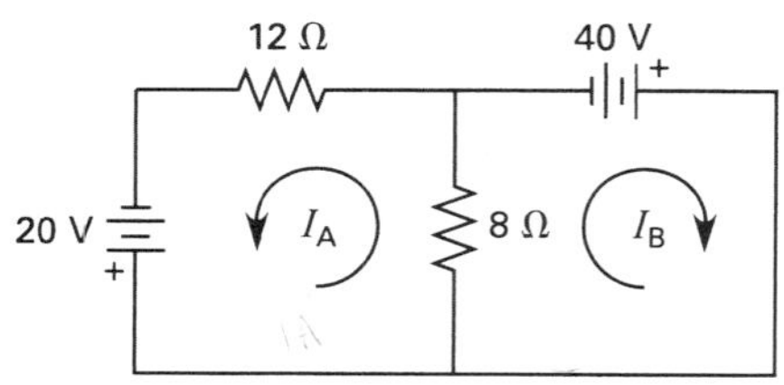

Most nearly, the current through the 40 V battery is

(A) 1.3 A
(B) 2.9 A
(C) 6.7 A
(D) 13 A

8. A circuit containing four resistors is shown.

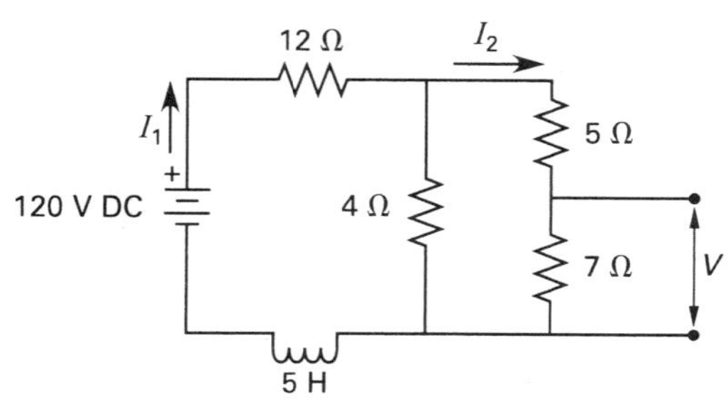

Most nearly, the voltage across the 7 Ω resistor is

(A) 7.0 V
(B) 11 V
(C) 14 V
(D) 24 V

9. A circuit containing three resistors is shown.

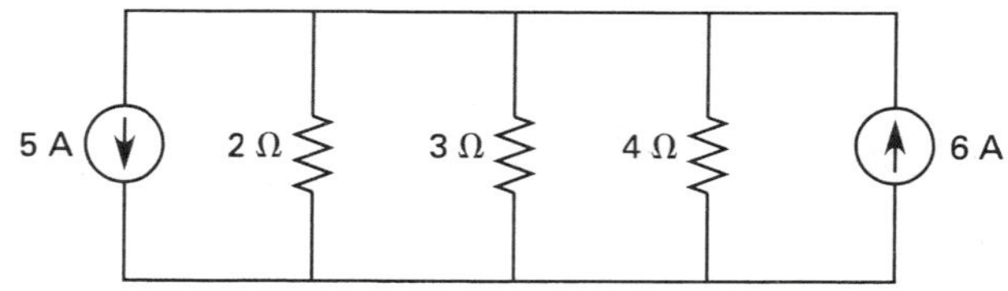

Most nearly, the current passing through the 2 Ω resistor is

(A) 0.19 A
(B) 0.46 A
(C) 0.82 A
(D) 1.1 A

10. A current source is placed in parallel with two conductances.

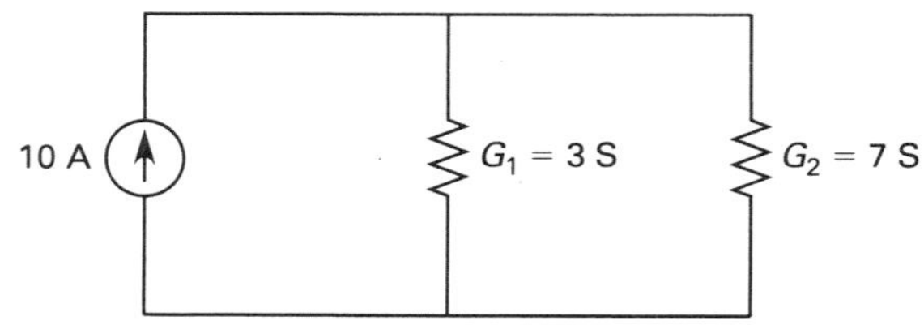

Most nearly, the power absorbed by the 3 S conductance is

(A) 1 W
(B) 3 W
(C) 5 W
(D) 7 W

11. Two voltage sources and two resistors are combined as shown. The resulting current is 0.1 A, clockwise.

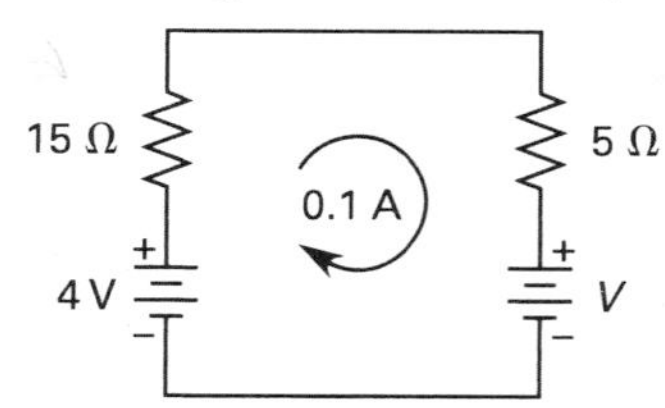

Most nearly, the unknown voltage, V, is

(A) 1.0 V
(B) 2.0 V
(C) 4.0 V
(D) 10 V

12. The equivalent circuit for a power supply consists of an ideal 10 V voltage in series with a 4 Ω resistance. The power supply is connected to a 1 Ω load.

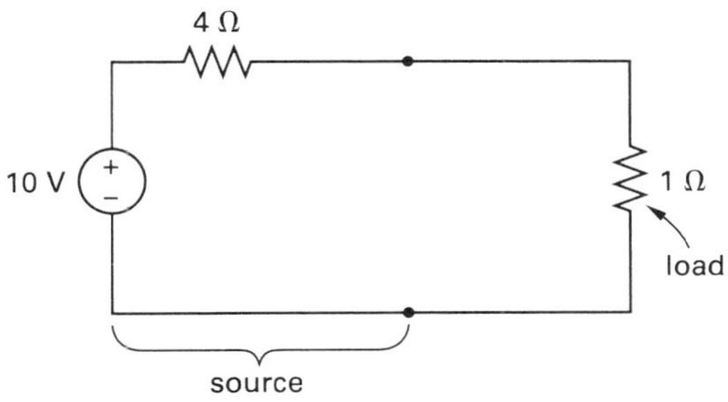

Most nearly, how much would the voltage across the 1 Ω load resistance have to change in order to dissipate maximum power?

(A) 1 V

(B) 2 V

(C) 3 V

(D) 5 V

SOLUTIONS

1. The current through the 6 Ω resistor can be found through superposition of the current due to the voltage source and the current due to the current source. First, replace the current source with an open circuit as shown.

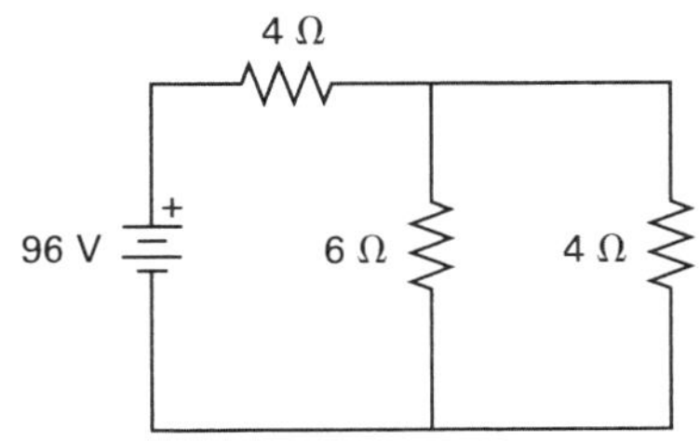

Calculate the total resistance. The 6 Ω resistor and one of the 4 Ω resistors are in parallel, so

$$\begin{aligned} R_{\text{total}} &= R_1 + \frac{R_2 R_3}{R_2 + R_3} \\ &= 4\ \Omega + \frac{(4\ \Omega)(6\ \Omega)}{4\ \Omega + 6\ \Omega} \\ &= 6.4\ \Omega \end{aligned}$$

The total current is

$$I_{\text{total}} = \frac{V_{\text{total}}}{R_{\text{total}}} = \frac{96\ \text{V}}{6.4\ \Omega} = 15\ \text{A}$$

The current through the 6 Ω resistor due to the voltage source is

$$I_{2,\text{voltage}} = I_{\text{total}}\left(\frac{R_3}{R_3 + R_2}\right) = (15\ \text{A})\left(\frac{4\ \Omega}{4\ \Omega + 6\ \Omega}\right) = 6\ \text{A}$$

Next, in the original circuit, replace the voltage source with a short circuit.

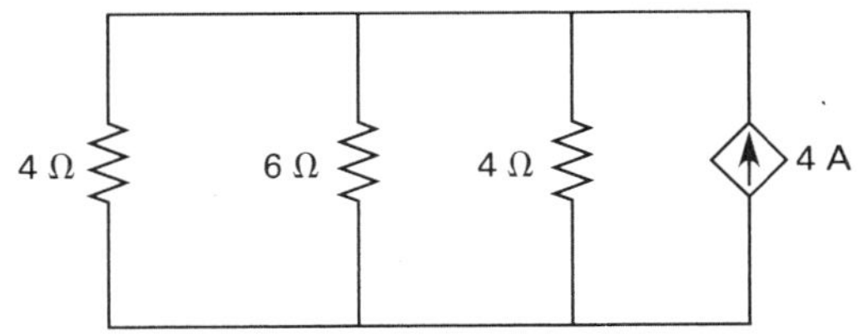

The two 4 Ω resistors in parallel are equivalent to a 2 Ω resistor, so the current through the 6 Ω resistor due to the current source is

$$I_{2,\text{current}} = I\left(\frac{R_{13}}{R_2 + R_{13}}\right) = (4\ \text{A})\left(\frac{2\ \Omega}{6\ \Omega + 2\ \Omega}\right) = 1\ \text{A}$$

The total current through the 6 Ω resistor is

$$6\ \text{A} + 1\ \text{A} = 7\ \text{A} \quad (7.0\ \text{A})$$

The answer is (A).

2. The open-circuit Thevenin equivalent voltage across terminals A and B is the voltage across the vertical 3 kΩ resistor.

$$\begin{aligned} V_{\text{Th}} &= V_{\text{AB}} = RI \\ &= \left((3\ \text{k}\Omega)\left(1000\ \frac{\Omega}{\text{k}\Omega}\right)\right)\left(\frac{30\ \text{mA}}{1000\ \frac{\text{mA}}{\text{A}}}\right) \\ &= 90\ \text{V} \end{aligned}$$

The Thevenin equivalent resistance can be determined by reducing the current source to zero. The left leg becomes an open circuit. The resistance across terminals A and B is

$$\begin{aligned} R_{\text{Th}} &= 3\ \text{k}\Omega + 3\ \text{k}\Omega \\ &= 6\ \text{k}\Omega \end{aligned}$$

The Thevenin equivalent circuit across terminals A and B is a series voltage source of 90 V and a resistance of 6 kΩ.

The answer is (D).

3. The problem can be simplified by converting the current source and the 3 Ω resistor to a Thevenin equivalent.

$$\begin{aligned} V_{\text{oc}} &= R_{\text{Th}} I_{\text{short circuit}} = (3\ \Omega)(4\ \text{A}) \\ &= 12\ \text{V} \end{aligned}$$

The simplified circuit is

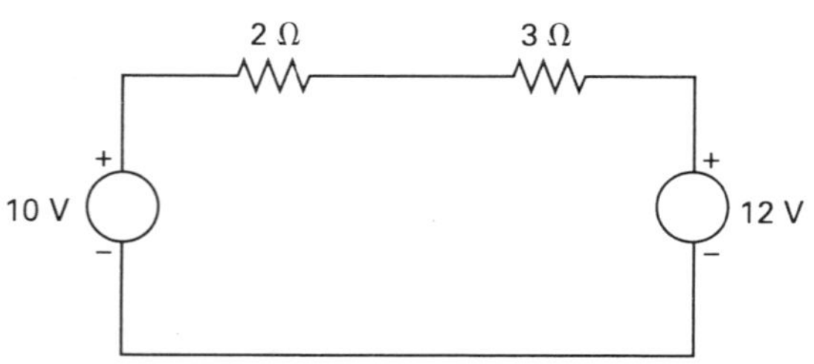

The current through the 2 Ω resistor is

$$i_{2\Omega} = \frac{V}{R} = \frac{12\ \text{V} - 10\ \text{V}}{2\ \Omega + 3\ \Omega} = 0.4\ \text{A}$$

The 12 V source is greater than the 10 V source, so the current of 0.4 A is from right to left.

The answer is (A).

4. The maximum (full-scale) voltage drop across the meter is $V_{\text{fs}} = 20$ V. The maximum current through the meter is $I_{\text{fs}} = 50\ \mu$A. The total internal resistance of the voltmeter is

$$\begin{aligned} R_{\text{total}} &= \frac{V_{\text{fs}}}{I_{\text{fs}}} = \frac{(20\ \text{V})\left(10^6\ \frac{\mu\text{A}}{\text{A}}\right)}{(50\ \mu\text{A})\left(1000\ \frac{\Omega}{\text{k}\Omega}\right)} \\ &= 400\ \text{k}\Omega \end{aligned}$$

The total resistance is the sum of the meter and series resistances.

$$R_{\text{total}} = R_{\text{meter}} + R_{\text{series}}$$

The series resistance is

$$\begin{aligned} R_{\text{series}} &= R_{\text{total}} - R_{\text{meter}} \\ &= 400\ \text{k}\Omega - 5\ \text{k}\Omega \\ &= 395\ \text{k}\Omega \quad (400\ \text{k}\Omega) \end{aligned}$$

The answer is (C).

5. Because the voltmeter is connected in parallel with the load resistance, its internal resistance must be very high so that it will not draw enough current to significantly affect the current flowing through the measurement leads.

The answer is (D).

6. Maximum power transfer occurs when the load resistance equals the source resistance. Find the Thevenin equivalent resistance for the circuit to the left of terminals A and B. The Thevenin equivalent resistance is found by removing the voltage source and finding the equivalent resistance across terminals A and B with R_L disconnected. The equivalent resistance is a series combination of the 2 Ω resistance and the parallel combination of the 15 Ω and 10 Ω resistances.

$$\begin{aligned} R_S &= R_3 + R_P \\ &= R_3 + \frac{R_1 R_2}{R_1 + R_2} \\ &= 2\ \Omega + \frac{(10\ \Omega)(15\ \Omega)}{10\ \Omega + 15\ \Omega} \\ &= 8.0\ \Omega \end{aligned}$$

The answer is (B).

7. Use the current-loop method to solve for the unknown currents. Use Kirchhoff's voltage law to get an equation for each loop.

$$\begin{aligned}\sum V &= \sum IR \\ 20\ \text{V} &= (I_\text{A} + I_\text{B})(8\ \Omega) + I_\text{A}(12\ \Omega) \\ 40\ \text{V} &= (I_\text{A} + I_\text{B})(8\ \Omega)\end{aligned}$$

Rearrange the second equation to solve for I_A.

$$I_\text{A} = \frac{40\ \text{V}}{8\ \Omega} - I_\text{B} = 5\ \text{A} - I_\text{B}$$

Substitute this value for I_A in the first equation, then solve for I_B.

$$\begin{aligned}20\ \text{V} &= \big((5\ \text{A} - I_\text{B}) + I_\text{B}\big)(8\ \Omega) + (5\ \text{A} - I_\text{B})(12\ \Omega) \\ I_\text{B} &= 6.67\ \text{A} \quad (6.7\ \text{A})\end{aligned}$$

The answer is (C).

8. Determine the equivalent resistance of the two parallel legs, one containing a 4 Ω resistor and the other containing a 5 Ω resistor and a 7 Ω resistor in series.

$$\begin{aligned}R_\text{parallel} &= \frac{1}{\dfrac{1}{R_1} + \dfrac{1}{R_S}} = \frac{1}{\dfrac{1}{R_1} + \dfrac{1}{R_2 + R_3}} \\ &= \frac{1}{\dfrac{1}{4\ \Omega} + \dfrac{1}{5\ \Omega + 7\ \Omega}} \\ &= 3\ \Omega\end{aligned}$$

The 12 Ω resistor and this combination of resistors are connected in series. Because the voltage source is DC, the inductor contributes no resistance. The total equivalent resistance of the circuit is

$$\begin{aligned}R_\text{eq} &= 12\ \Omega + 3\ \Omega \\ &= 15\ \Omega\end{aligned}$$

Use Ohm's law to calculate the currents and voltages through individual elements. The current I_1 is

$$I_1 = \frac{V}{R_\text{eq}} = \frac{120\ \text{V}}{15\ \Omega} = 8\ \text{A}$$

The voltage across the parallel section is

$$V_\text{parallel} = I_1 R_\text{parallel} = (8\ \text{A})(3\ \Omega) = 24\ \text{V}$$

$$I_2 = \frac{V_\text{parallel}}{R} = \frac{24\ \text{V}}{5\ \Omega + 7\ \Omega} = 2\ \text{A}$$

Use Ohm's law to calculate the voltage across the 7 Ω resistor.

$$V = I_2 R = (2\ \text{A})(7\ \Omega) = 14\ \text{V}$$

The answer is (C).

9. The two current sources are connected to the same points, so they can be combined. Taking an upward current as positive, the net current is

$$6\ \text{A} - 5\ \text{A} = 1\ \text{A}$$

Draw an equivalent circuit.

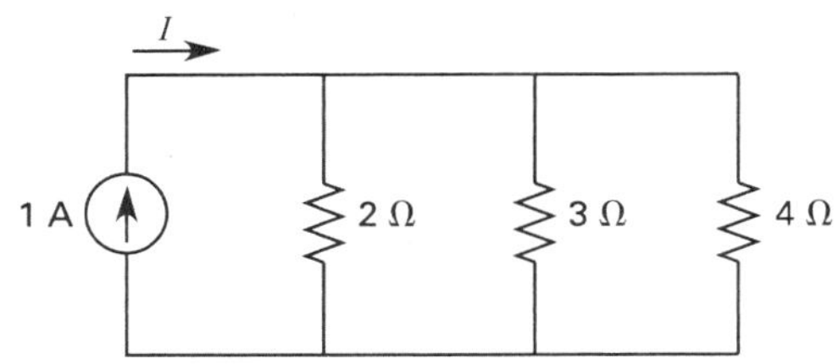

The three resistors are in parallel, so the equivalent resistance of the circuit is

$$\begin{aligned}R_\text{eq} &= \frac{1}{\dfrac{1}{R_1} + \dfrac{1}{R_2} + \dfrac{1}{R_3}} = \frac{1}{\dfrac{1}{2\ \Omega} + \dfrac{1}{3\ \Omega} + \dfrac{1}{4\ \Omega}} \\ &= 0.923\ \Omega\end{aligned}$$

The voltage across all the resistors is

$$V = IR_\text{eq} = (1\ \text{A})(0.923\ \Omega) = 0.923\ \text{V}$$

The current through the 2 Ω resistor is

$$I_{2\ \Omega} = \frac{V}{R} = \frac{0.923\ \text{V}}{2\ \Omega} = 0.462\ \text{A} \quad (0.46\ \text{A})$$

The concept of a current divider could also be used to solve this problem.

The answer is (B).

10. The conductances are in parallel, so the total conductance is their sum.

$$G_\text{total} = G_1 + G_2 = 3\ \text{S} + 7\ \text{S} = 10\ \text{S}$$

The voltage across the conductances is

$$\begin{aligned}V &= IR_\text{total} = \frac{I}{G_\text{total}} = \frac{10\ \text{A}}{10\ \text{S}} \\ &= 1\ \text{V}\end{aligned}$$

The current through the G_1 conductance is

$$\begin{aligned} I_1 &= \frac{V}{R_1} = VG_1 = (1\ \text{V})(3\ \text{S}) \\ &= 3\ \text{A} \end{aligned}$$

The power dissipated by the G_1 conductance is

$$\begin{aligned} P_1 &= I_1^2 R_1 = \frac{I_1^2}{G_1} = \frac{(3\ \text{A})^2}{3\ \text{S}} \\ &= 3\ \text{W} \end{aligned}$$

The answer is (B).

11. Going clockwise around the loop, the loop equation is

$$\begin{aligned} \sum V &= I \sum R \\ 4\ \text{V} - V &= (0.1\ \text{A})(15\ \Omega + 5\ \Omega) \\ &= 2\ \text{V} \end{aligned}$$

Solving for the unknown voltage gives

$$V = 4\ \text{V} - 2\ \text{V} = 2\ \text{V}$$

The answer is (B).

12. Using a voltage divider concept to calculate the load voltage gives

$$V_{\text{actual}} = \left(\frac{R_L}{R_L + R_{\text{Th}}} \right) V_{\text{Th}} = \left(\frac{1\ \Omega}{1\ \Omega + 4\ \Omega} \right) (10\ \text{V}) = 2\ \text{V}$$

In the maximum power transfer configuration, the load resistance would be equal to the Thevenin source resistance, 4 Ω. The voltage across the load resistance would then be

$$V_{\text{max power}} = \left(\frac{R_L}{R_L + R_{\text{Th}}} \right) V_{\text{Th}} = \left(\frac{4\ \Omega}{4\ \Omega + 4\ \Omega} \right) (10\ \text{V}) = 5\ \text{V}$$

The difference in voltage across the load resistance is

$$V_{\text{difference}} = V_{\text{max power}} - V_{\text{actual}} = 5\ \text{V} - 2\ \text{V} = 3\ \text{V}$$

The answer is (C).

24 Alternating-Current Circuits

PRACTICE PROBLEMS

1. A standby generator provides backup power to a process fed by a 600 A power distribution panel. The process uses four 100 hp pumps each having an efficiency of 85%. Four three-phase motors driving the pumps operate at 480 V and have a power factor of 0.8. Most nearly, the minimum size generator needed is

(A) 110 kVA

(B) 350 kVA

(C) 440 kVA

(D) 500 kVA

2. A circuit consists of a pure resistance in series with an unknown impedance. When the circuit is connected to a 480 V source, the pure resistance draws 10 kW, and the unknown impedance draws 40 kW with a 0.8 power factor.

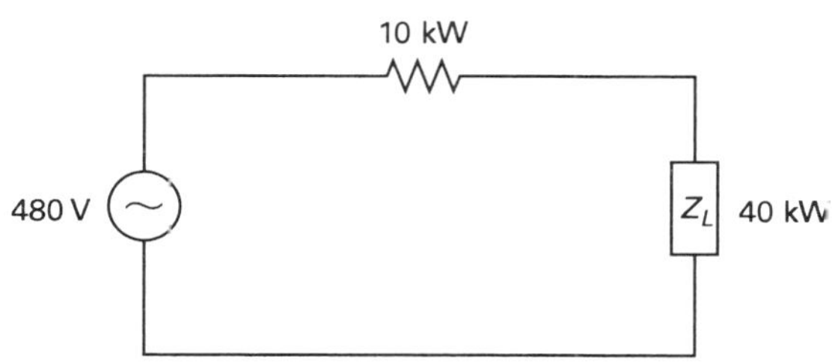

What is the power factor for the overall circuit?

(A) 0.50

(B) 0.80

(C) 0.86

(D) 0.92

3. An ideal single-phase transformer is rated as 440 V:110 V. The primary voltage is 440 V, and the secondary load is 5.5 Ω resistive. What are the actual primary and secondary currents?

(A) 5 A primary, 20 A secondary

(B) 5 A primary, 80 A secondary

(C) 20 A primary, 5 A secondary

(D) 80 A primary, 20 A secondary

4. Two single-phase voltage sources, v_1 and v_2, are connected in series.

$$v_1 = 20\ \sin(377t + 25°)$$
$$v_2 = 30\ \cos(377t - 15°)$$

What is the total voltage $v_3 = v_1 + v_2$?

(A) 50.0 tan(377t + 10°)

(B) 45.5 sin(377t + 55.3°)

(C) 10.0 tan(377t + 10°)

(D) 45.5 cos(377t + 55.3°)

5. An ideal transformer has primary and secondary windings designated 1 and 2, respectively. Variable I is current, V is applied voltage, Z is impedance, and N is the number of turns. Which of the following is true?

(A) $I_1 = I_2$

(B) $Z_2 = V_1/I_1$

(C) $I_1/I_2 = N_2/N_1$

(D) $I_1V_1N_1 = I_2V_2N_2$

6. The ideal transformer shown has a turns ratio of 5. The primary winding is connected to a 1200 V AC current. The primary current is 4.8 A. The secondary winding is connected to a 10 Ω ideal resistor.

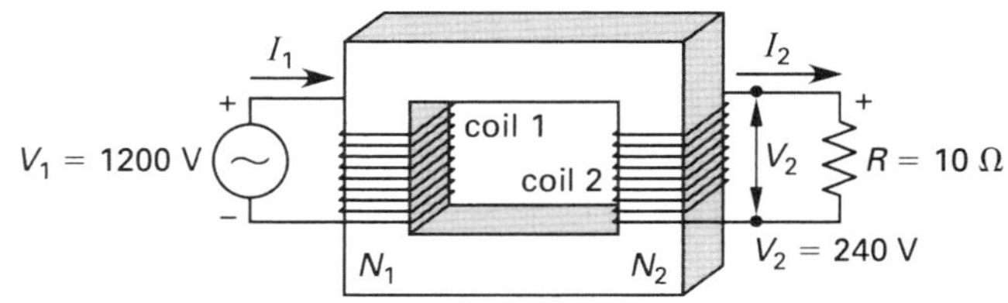

Most nearly, the current flowing in the secondary is

(A) 12 A

(B) 24 A

(C) 29 A

(D) 120 A

7. An ideal transformer has a turns ratio of 5. The primary winding is connected to a 1200 V AC current. The secondary winding is connected to a 10 Ω ideal resistor. The voltage across the resistor is measured to be 240 V.

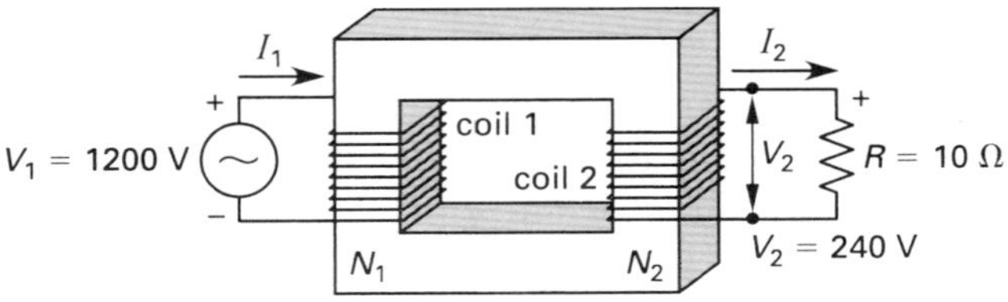

Most nearly, the primary current is

(A) 1.9 A

(B) 4.8 A

(C) 24 A

(D) 120 A

8. A DC generator produces 200 V at 1800 rpm. The speed is decreased to 1000 rpm. Armature resistance is negligible. Most nearly, the new voltage generated is

(A) 100 V

(B) 110 V

(C) 200 V

(D) 360 V

9. The voltage for a waveform is 2 V + (5 V)(cos $2\pi t$). Most nearly, the effective value of this waveform is

(A) 3.0 V

(B) 4.1 V

(C) 5.5 V

(D) 7.0 V

10. A waveform, $f(t)$, is represented as

$$f(t) = \cos\left(\omega t + \frac{\pi}{2}\right)$$

Which of the following represents the same waveform?

(A) $\cos \omega t$

(B) $\cos(\omega t - (\pi/2))$

(C) $-\sin \omega t$

(D) $\sin \omega t$

11. What is the following voltage expressed in phasor form?

$$(170 \text{ V})\cos\left(100t + \frac{\pi}{3}\right)$$

(A) $120 \text{ V}\angle\frac{\pi}{3}$ rad

(B) $120 \text{ V}\angle 60°$

(C) $170 \text{ V}\angle\frac{\pi}{3}$ rad

(D) $170 \text{ V}\angle 60°$

12. The series *RLC* circuit shown is connected to a voltage source having a frequency of 400 rad/s.

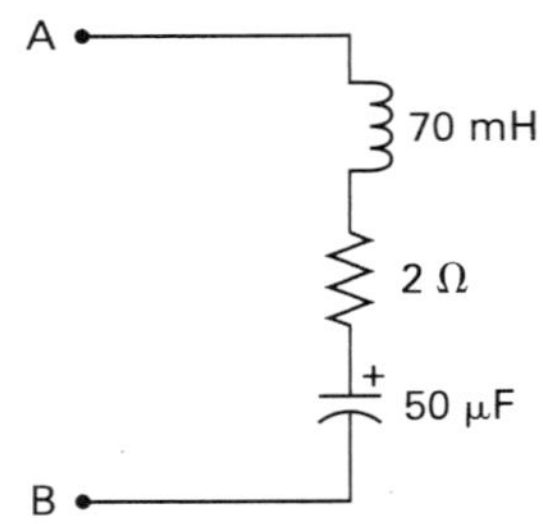

Most nearly, the resulting impedance between terminals A and B is

(A) $11.0\ \Omega\angle{-79.6°}$

(B) $15.6\ \Omega\angle{-84.8°}$

(C) $15.6\ \Omega\angle 84.8°$

(D) $22.1\ \Omega\angle{-84.8°}$

13. A generator develops a voltage of $120 \text{ V}\angle 0°$ and has an internal impedance of $20 + j20\ \Omega$. The generator is connected to a load of $30\ \Omega + jX_{\text{load}}$ by means of a transmission line with an impedance of $R_{\text{trans}} - j10\ \Omega$.

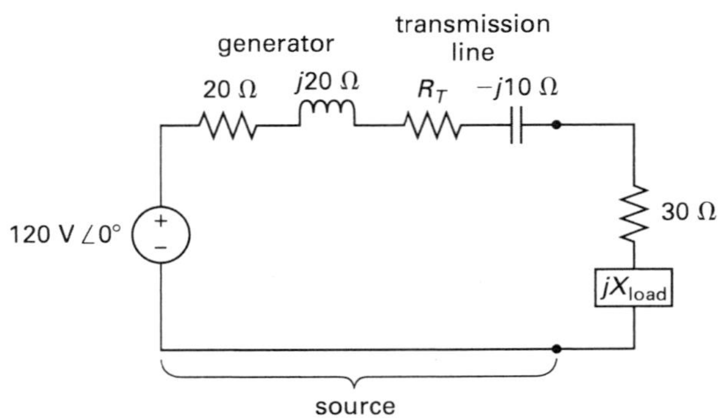

Most nearly, what values of R_T and X_{load} will cause maximum power to be transferred?

(A) $R_{trans} = 10\ \Omega$, $X_{load} = -10\ \Omega$

(B) $R_{trans} = 10\ \Omega$, $X_{load} = 10\ \Omega$

(C) $R_{trans} = 20\ \Omega$, $X_{load} = -10\ \Omega$

(D) $R_{trans} = 30\ \Omega$, $X_{load} = 20\ \Omega$

14. A sinusoidal voltage source has a voltage maximum of 120 V. The resulting sinusoidal current has a maximum of 4 A. The real power drawn is 192 W. Most nearly, the power factor is

(A) 0.4

(B) 0.5

(C) 0.6

(D) 0.8

15. A sinusoidal voltage source has voltage maximum of 120 V and current maximum of 4 A. The power factor is 0.85 leading. Most nearly, the reactive power is

(A) −126 VAR

(B) −63.2 VAR

(C) 63.2 VAR

(D) 126 VAR

16. A sinusoidal source produces an effective voltage of 100 V. The source contains an internal impedance of $4\ \Omega - j20\ \Omega$ and is connected to a $6\ \Omega$ load.

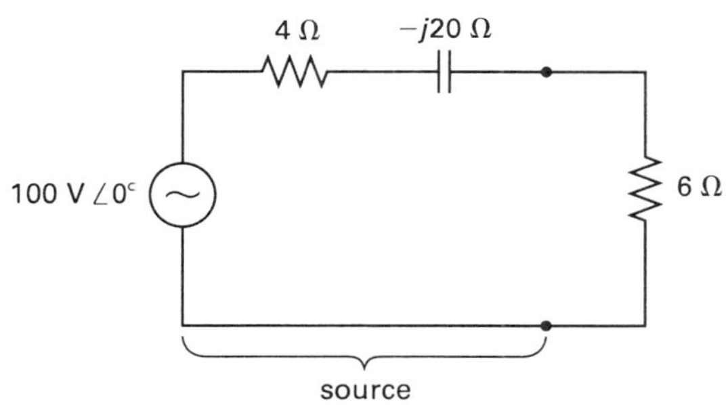

Most nearly, the real power dissipated in the load resistance is

(A) 60 W

(B) 80 W

(C) 120 W

(D) 240 W

17. An ideal transfer contains 100 primary coil turns and 400 secondary coil turns. The impedance measured at the primary terminals of the transformer is $100 - j50\ \Omega$. Most nearly, the impedance in the secondary circuit is

(A) $100 - j50\ \Omega$

(B) $400 - j200\ \Omega$

(C) $1600 - j800\ \Omega$

(D) $1600 + j800\ \Omega$

18. A 240 V alternating source at 60 Hz is connected to a series-*RLC* circuit as shown.

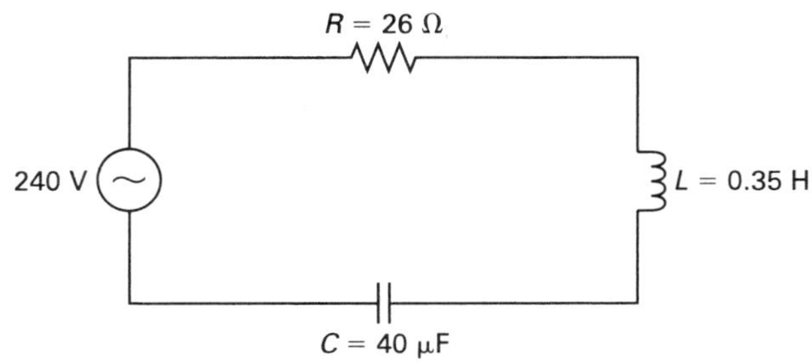

What is most nearly the total reactance of the circuit?

(A) 66 Ω

(B) 130 Ω

(C) 150 Ω

(D) 200 Ω

Circuit Analysis/ Linear Systems

SOLUTIONS

1. Use the efficiency to calculate the power needed to provide 100 horsepower to the pumps.

$$\eta = \frac{P_L}{P_S}$$

$$\begin{aligned} P_S &= \frac{P_L}{\eta} = \frac{(100\ \text{hp})\left(0.746\ \frac{\text{kW}}{\text{hp}}\right)}{0.85} \\ &= 87.76\ \text{kW} \end{aligned}$$

Use the equation for power to find the current drawn by each motor. (Because the motors are three-phase, the voltage is multiplied by $\sqrt{3}$.)

$$\begin{aligned} P &= V_{\text{motor}} I_{\text{motor}} \cos\theta \\ P &= V_{\text{motor}} I_{\text{motor}}(\text{pf}) \\ I_{\text{motor}} &= \frac{P}{V_{\text{motor}}(\text{pf})} \\ &= \frac{(87.76\ \text{kW})\left(1000\ \frac{\text{W}}{\text{kW}}\right)}{\sqrt{3}\,(480\ \text{V})(0.8)} \\ &= 131.9\ \text{A} \quad [\text{each motor}] \end{aligned}$$

The generator must be capable of supplying 131.9 A at 480 V to each of the four motors. The minimum size generator needed is

$$\begin{aligned} S_{\text{generator}} &= \frac{(4\ \text{motors})\sqrt{3}\,(480\ \text{V})\left(131.9\ \frac{\text{A}}{\text{motor}}\right)}{1000\ \frac{\text{VA}}{\text{kVA}}} \\ &= 438.8\ \text{kVA} \quad (440\ \text{kVA}) \end{aligned}$$

The answer is (C).

2. The power triangle for the unknown impedance is

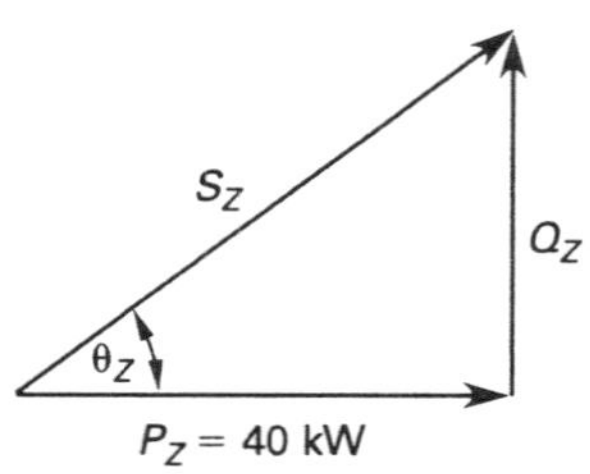

$$\begin{aligned} \text{pf} &= \cos\theta_Z \\ \theta_Z &= \arccos \text{pf} = \arccos 0.8 \\ &= 36.87^\circ \\ Q_Z &= P_Z \tan\theta_Z = (40\,\text{kW})\tan 36.87^\circ \\ &= 30\ \text{kVAR} \end{aligned}$$

For the overall system, the power triangle is

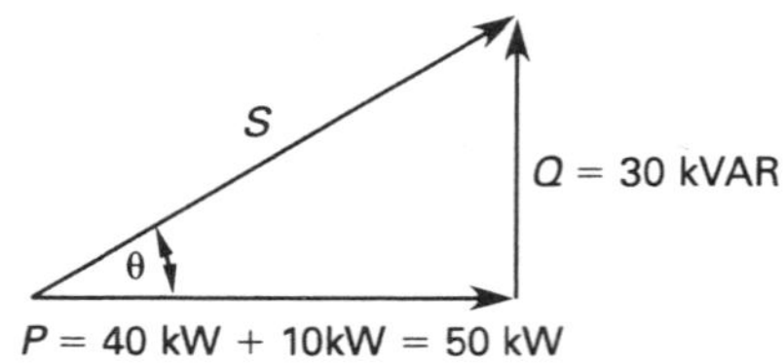

$$\begin{aligned} \theta &= \arctan\frac{Q}{P} = \arctan\frac{30\ \text{kVAR}}{50\ \text{kW}} \\ &= 31^\circ \\ \text{pf} &= \cos\theta = \cos 31^\circ \\ &= 0.86 \end{aligned}$$

The answer is (C).

3. The transformer is ideal, so its turns ratio is

$$\begin{aligned} a &= \left|\frac{\mathbf{V}_P}{\mathbf{V}_S}\right| \\ &= \frac{440\ \text{V}}{110\ \text{V}} \\ &= 4 \end{aligned}$$

The secondary current is

$$I_S = \frac{V_S}{R_S} = \frac{110\ \text{V}}{5.5\ \Omega} = 20\ \text{A}$$

For an ideal transformer,

$$\begin{aligned} a &= \left|\frac{\mathbf{I}_S}{\mathbf{I}_P}\right| \\ I_P &= \frac{I_S}{a} = \frac{20\ \text{A}}{4} \\ &= 5\ \text{A} \end{aligned}$$

The answer is (A).

4. Express v_2 as a sine function. The sine and cosine functions differ by 90°.

$$\begin{aligned} v_2 &= 30\cos(377t - 15^\circ) \\ &= 30\sin(377t - 15^\circ + 90^\circ) \\ &= 30\sin(377t + 75^\circ) \end{aligned}$$

Convert v_1 and v_2 to phasor form.

$$\begin{aligned} v_1 &= 20(\cos 25° + j\ \sin 25°) \\ &= 18.126 + j8.452 \\ v_2 &= 30(\cos 75° + j\ \sin 75°) \\ &= 7.765 + j28.978 \end{aligned}$$

The phasor sum of v_1 and v_2 is

$$\begin{aligned} v_3 &= (18.126 + 7.765) + j(8.452 + 28.978) \\ &= 25.891 + j37.430 \end{aligned}$$

Convert v_3 back to the time domain.

$$\begin{aligned} r &= \sqrt{(25.891)^2 + (37.430)^2} = 45.51 \\ \phi &= \arctan \frac{37.430}{25.891} = 55.33° \\ v_3 &= r\sin(\omega t + \phi) \\ &= 45.51 \sin(377t + 55.33°) \end{aligned}$$

The answer is (B).

5. An ideal transformer is lossless. All the power drawn by the primary winding is transferred to the secondary winding.

$$I_1 V_1 = I_2 V_2$$

The ratio of common terms defines the turns ratio.

$$a = \frac{V_1}{V_2} = \frac{I_2}{I_1} = \frac{N_1}{N_2}$$

The inverse of the turns ratio is

$$\frac{1}{a} = \frac{N_2}{N_1} = \frac{I_1}{I_2} = \frac{V_2}{V_1}$$

The answer is (C).

6. Use the turns ratio to calculate the current flowing in the secondary winding.

$$\begin{aligned} a &= \frac{I_2}{I_1} \\ I_2 &= aI_1 \\ &= (5)(4.8\ \text{A}) \\ &= 24\ \text{A} \end{aligned}$$

The answer is (B).

7. The current flowing through coil 2 is

$$I_2 = \frac{V_2}{R_2} = \frac{240\ \text{V}}{10\ \Omega} = 24\ \text{A}$$

Calculate the primary current from the turns ratio.

$$\begin{aligned} a &= \frac{I_2}{I_1} \\ I_1 &= \frac{I_2}{a} = \frac{24\ \text{A}}{5} \\ &= 4.8\ \text{A} \end{aligned}$$

The answer is (B).

8. The operating voltage is proportional to the rotational speed.

$$\begin{aligned} n_1/n_2 &= \left|\frac{\mathbf{V}_1}{\mathbf{V}_2}\right| \\ V_2 &= \left(\frac{n_2}{n_1}\right) V_1 \\ &= \left(\frac{1000\ \frac{\text{rev}}{\text{min}}}{1800\ \frac{\text{rev}}{\text{min}}}\right)(200\ \text{V}) \\ &= 111\ \text{V} \quad (110\ \text{V}) \end{aligned}$$

The answer is (B).

9. The effective voltage can be found in polar form.

$$V \equiv V_{\text{eff}} \angle \phi \equiv \left(\frac{V_{\max}}{\sqrt{2}}\right) \angle \phi = \frac{5\ \text{V}}{\sqrt{2}}$$

The effective value of the waveform is

$$\begin{aligned} X_{\text{rms}} &= \sqrt{X_{\text{dc}}^2 + \sum_{n=1}^{\infty} X_n^2} \\ V_{\text{rms}} &= \sqrt{V_{\text{dc}}^2 + \sum_{n=1}^{\infty} V_n^2} \\ &= \sqrt{(2\ \text{V})^2 + \left(\frac{5\ \text{V}}{\sqrt{2}}\right)^2} \\ &= 4.06\ \text{V} \quad (4.1\ \text{V}) \end{aligned}$$

The effective value can also be found in trigonometric form. Use the equation for frequency to find the period of the waveform.

$$\begin{aligned} f &= 1/T = \omega/2\pi \\ T &= \frac{2\pi}{\omega} = \frac{2\pi \text{ rad}}{2\pi \ \frac{\text{rad}}{\text{s}}} \\ &= 1 \text{ s} \end{aligned}$$

The effective value of the waveform is

$$X_{\text{eff}} = X_{\text{rms}} = \left[(1/T) \int_0^T x^2(t)\, dt \right]^{1/2}$$

$$\begin{aligned} V_{\text{eff}} = V_{\text{rms}} &= \sqrt{\frac{1}{1 \text{ s}} \int_0^T (2 \text{ V} + (5 \text{ V}) \cos 2\pi t)^2 \, dt} \\ &= \sqrt{\left(\frac{1}{1 \text{ s}}\right)\left(\int_{0 \text{ s}}^{1 \text{ s}} (2 \text{ V})^2 \, dt + \int_{0 \text{ s}}^{1 \text{ s}} ((5 \text{ V}) \cos 2\pi t)^2 \, dt\right)} \\ &= \sqrt{\left(\frac{1}{1 \text{ s}}\right)\left((4 \text{ V}^2)(1 \text{ s}) + (25 \text{ V}^2) \times \left(\frac{1 \text{ s}}{2} + \frac{(\sin 4\pi)(1 \text{ s})}{8\pi} - \frac{0 \text{ s}}{2} - \frac{(\sin 4\pi)(0 \text{ s})}{8\pi}\right)\right)} \\ &= \sqrt{4 \text{ V}^2 + \frac{25 \text{ V}^2}{2}} \\ &= 4.06 \text{ V} \quad (4.1 \text{ V}) \end{aligned}$$

The answer is (B).

10. According to sine-cosine relationships,

$$\sin \omega t = \cos\left(\omega t - \frac{\pi}{2}\right) = -\cos\left(\omega t + \frac{\pi}{2}\right)$$

Negating all terms,

$$-\sin \omega t = -\cos\left(\omega t - \frac{\pi}{2}\right) = \cos\left(\omega t + \frac{\pi}{2}\right)$$

The answer is (C).

11. When the voltage is expressed in phasor form, is magnitude is the effective value. Using the equation for the effective value of a sinusoid, the effective voltage is

$$\begin{aligned} X_{\text{eff}} &= X_{\text{rms}} = X_{\text{max}}/\sqrt{2} \\ V_{\text{eff}} &= V_{\text{rms}} = \frac{V_{\text{max}}}{\sqrt{2}} = \frac{170 \text{ V}}{\sqrt{2}} \\ &= 120.2 \text{ V} \quad (120 \text{ V}) \end{aligned}$$

In phasor form, the phase angle is expressed in degrees.

$$\phi = \left(\frac{\pi}{3} \text{ rad}\right)\left(\frac{180°}{\pi \text{ rad}}\right) = 60°$$

The phasor form of the voltage is

$$120 \text{ V}\angle 60°$$

The answer is (B).

12. The impedance of the resistor is 2 Ω. The impedance of the capacitor is

$$\begin{aligned} Z_C &= \frac{1}{j\omega C} \\ &= \frac{10^6 \ \frac{\mu\text{F}}{\text{F}}}{j\left(400 \ \frac{\text{rad}}{\text{s}}\right)(50 \ \mu\text{F})} \\ &= -j50 \ \Omega \end{aligned}$$

The impedance of the inductor is

$$\begin{aligned} Z_L &= j\omega L \\ &= \frac{j\left(400 \ \frac{\text{rad}}{\text{s}}\right)(70 \text{ mH})}{10^3 \ \frac{\text{mH}}{\text{H}}} \\ &= j28 \ \Omega \end{aligned}$$

The impedance is the sum of the impedances of the resistor, capacitor, and inductor.

$$\begin{aligned} Z &= Z_R + Z_C + Z_L = 2 \ \Omega - j50 \ \Omega + j28 \ \Omega \\ &= 2 \ \Omega - j22 \ \Omega \\ &= \sqrt{(2 \ \Omega)^2 + (-22 \ \Omega)^2} \ \angle \arctan \frac{-22 \ \Omega}{2 \ \Omega} \\ &= 22.09 \ \Omega\angle -84.8° \quad (22.1 \ \Omega\angle -84.8°) \end{aligned}$$

The answer is (D).

13. In the maximum power transfer arrangement, the resistance of the load is equal to the resistance of the source. The resistance of the source is equal to the sum of the resistances of the generator and the transmission line.

$$R_{\text{load}} = R_{\text{source}} = R_{\text{gen}} + R_{\text{trans}}$$

The resistance of the transmission line is

$$R_{\text{trans}} = R_{\text{load}} - R_{\text{gen}} = 30 \ \Omega - 20 \ \Omega = 10 \ \Omega$$

The reactance of the load is the negative of the reactance of the source, which is equal to the sum of the reactances of the generator and the transmission line.

$$\begin{aligned} X_{\text{load}} &= -X_{\text{source}} = -(X_{\text{gen}} + X_{\text{trans}}) \\ &= -(20\ \Omega - 10\ \Omega) \\ &= -10\ \Omega \end{aligned}$$

The answer is (A).

14. Use the equation for real power to find the power factor.

$$\begin{aligned} P &= \left(\tfrac{1}{2}\right) V_{\text{max}} I_{\text{max}} \cos\theta \\ \cos\theta &= \frac{2P}{V_{\text{max}} I_{\text{max}}} \\ \text{pf} &= \frac{(2)(192\ \text{W})}{(120\ \text{V})(4\ \text{A})} \\ &= 0.8 \end{aligned}$$

The answer is (D).

15. The power factor is leading, so the power angle, θ, is negative.

$$\theta = -\arccos(\text{pf}) = -\arccos 0.85 = -31.79°$$

The reactive power is

$$\begin{aligned} Q &= \left(\tfrac{1}{2}\right) V_{\text{max}} I_{\text{max}} \sin\theta \\ &= \left(\tfrac{1}{2}\right)(120\ \text{V})(4\ \text{A})\sin -31.79° \\ &= -126.4\ \text{VAR} \quad (-126\ \text{VAR}) \end{aligned}$$

The answer is (A).

16. From Ohm's law, the current in the circuit is

$$\begin{aligned} \mathbf{Z} &= \mathbf{V}/\mathbf{I} \\ I &= \frac{V_{\text{gen}}}{Z} = \frac{100\ \text{V}\angle 0°}{4\ \Omega + 6\ \Omega - j20\ \Omega} \\ &= \frac{100\ \text{V}\angle 0°}{10\ \Omega - j20\ \Omega} \\ |Z| &= \sqrt{(10\ \Omega)^2 + (-20\ \Omega)^2} = 22.36\ \Omega \\ \theta &= \arctan\frac{-20\ \Omega}{10\ \Omega} = -63.43° \\ Z &= 22.36\ \Omega\angle -63.43° \\ I &= \frac{V}{Z} = \frac{100\ \text{V}\angle 0°}{22.36\ \Omega\angle -63.43°} \\ &= 4.47\ \text{A}\angle 63.43° \end{aligned}$$

The voltage across the load is

$$\begin{aligned} V_R &= \left(\frac{R}{Z}\right) V_G = \left(\frac{6\ \Omega}{22.36\ \Omega\angle -63.43°}\right) 100\ \text{V}\angle 0° \\ &= 26.8\ \text{V}\angle 63.43° \end{aligned}$$

The voltage across the load and current are at the same phase angle, so the power angle, θ, is 0 and the power factor, pf, is 1.

The power may be determined in any of three ways.

$$\begin{aligned} P &= V_{\text{rms}} I_{\text{rms}} \cos\theta = (26.8\ \text{V})(4.47\ \text{A})(1) \\ &= 120\ \text{W} \\ P &= \frac{V_{\text{rms}}^2}{R} = \frac{(26.8\ \text{V})^2}{6\ \Omega} = 120\ \text{W} \\ P &= I_{\text{rms}}^2 R = (4.47\ \text{A})^2(6\ \Omega) = 120\ \text{W} \end{aligned}$$

The answer is (C).

17. The turns ratio is

$$a = N_1/N_2 = \frac{100\ \text{turns}}{400\ \text{turns}} = 0.25$$

The secondary impedance is

$$\begin{aligned} \mathbf{Z}_P &= a^2 \mathbf{Z}_S \\ \mathbf{Z}_S &= \frac{\mathbf{Z}_P}{a^2} = \frac{100 - j50\ \Omega}{(0.25)^2} \\ &= 1600 - j800\ \Omega \end{aligned}$$

The answer is (C).

18. The inductive reactance is

$$\begin{aligned} X_L &= \omega L = 2\pi f L = 2\pi(60\ \text{Hz})(0.35\ \text{H}) \\ &= 131.9\ \Omega \end{aligned}$$

The capacitive reactance is

$$\begin{aligned} X_C &= -\frac{1}{\omega C} = -\frac{1}{2\pi f C} \\ &= -\frac{1}{2\pi(60\ \text{Hz})(40\times 10^{-6}\ \text{F})} \\ &= -66.3\ \Omega \end{aligned}$$

The total reactance is

$$\begin{aligned} X_L + X_C &= 131.9\ \Omega + (-66.3\ \Omega) \\ &= 65.6\ \Omega \quad (66\ \Omega) \end{aligned}$$

The answer is (A).

25 Transient, Resonant, and Filter Circuits

PRACTICE PROBLEMS

1. The switch in the circuit shown has been open for a long time, and there is no current flowing through the inductor.

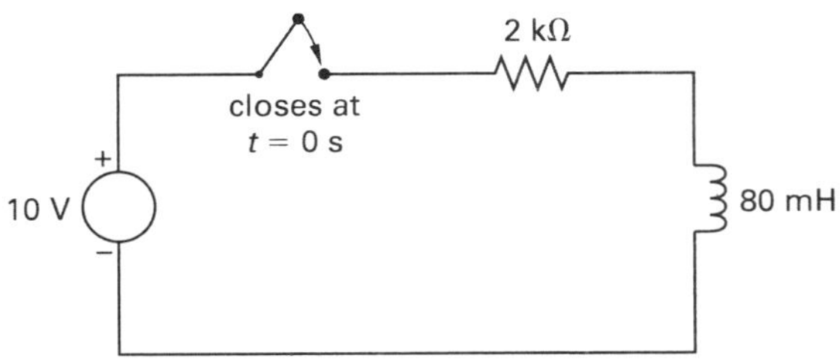

The switch closes at time $t = 0$ s. Most nearly, the current passing through the 80 mH inductor 60 μs after the switch is closed is

(A) 2.4 mA

(B) 3.2 mA

(C) 3.9 mA

(D) 5.0 mA

2. The circuit shown is to be used as a filter. The cutoff frequency each resistor-capacitor pair is 318 Hz.

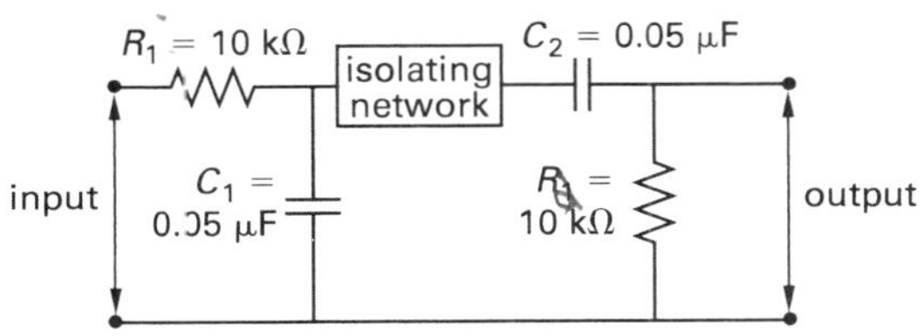

This circuit is a

(A) low-pass filter

(B) high-pass filter

(C) band-reject filter

(D) band-pass filter

3. The switch in the circuit shown has been open for a long time, and there is no initial charge on the capacitor.

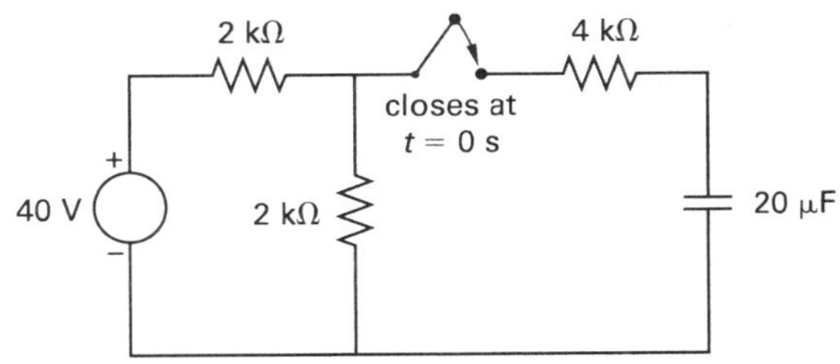

The switch is closed at time $t = 0$. Most nearly, the steady-state voltage across the capacitor after the switch is closed is

(A) 10 V

(B) 20 V

(C) 30 V

(D) 40 V

4. A voltage of $40 \sin 100t$ is impressed on the input of the filter circuit shown.

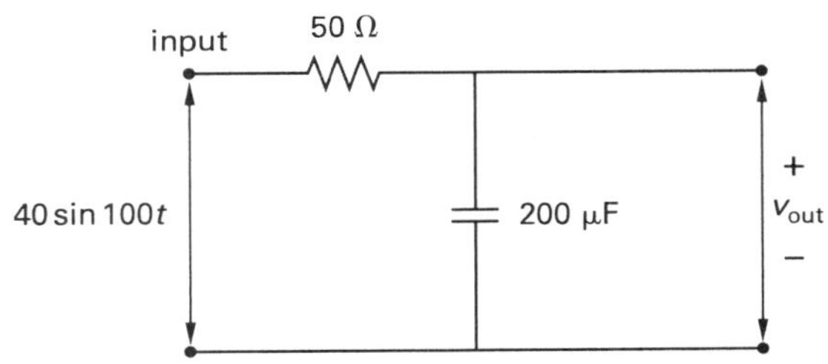

What is the amplitude of the output, v_{out}?

(A) 0 V

(B) 20 V

(C) 28 V

(D) 40 V

5. The exponentially decaying voltage given is applied to a digital circuit input.

$$v(t) = 20e^{-t/400 \times 10^{-3}\ \text{s}}$$

The circuit's trigger voltage is 2 V. Most nearly, the time needed for the decaying voltage to reach the trigger voltage is

(A) 200 ms
(B) 400 ms
(C) 920 ms
(D) 1800 ms

6. The switch in the circuit shown has been open for a long time, and there is no initial charge on the capacitor. The switch is closed at time $t = 0$ s. Most nearly, the capacitor will be charged to 80% of the battery voltage in

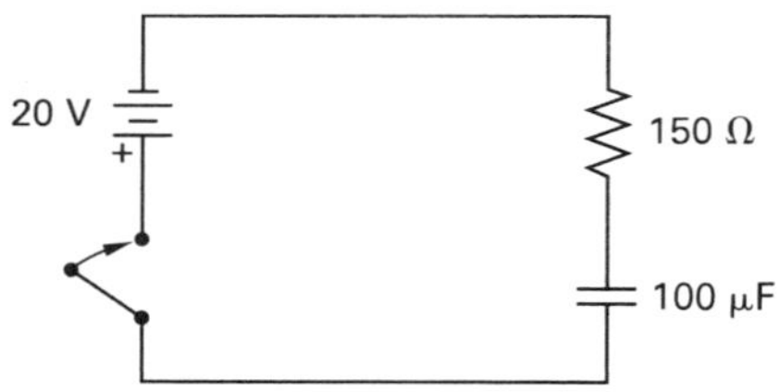

(A) 2.0 ms
(B) 10 ms
(C) 12 ms
(D) 24 ms

7. What is the resonant frequency of the circuit shown?

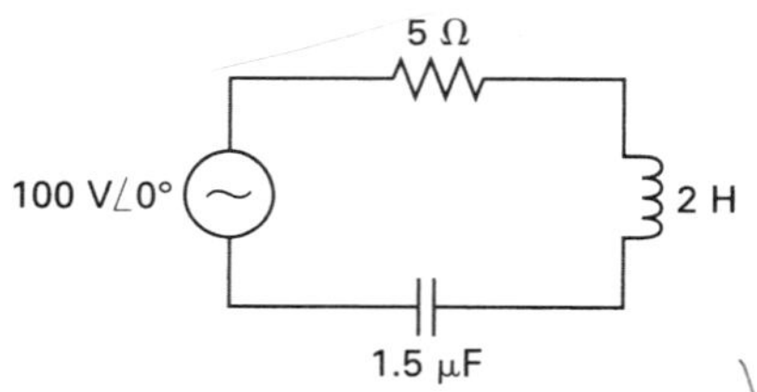

(A) 0 rad/s
(B) 92 rad/s
(C) 260 rad/s
(D) 580 rad/s

8. Most nearly, what is the resonant frequency of the circuit shown?

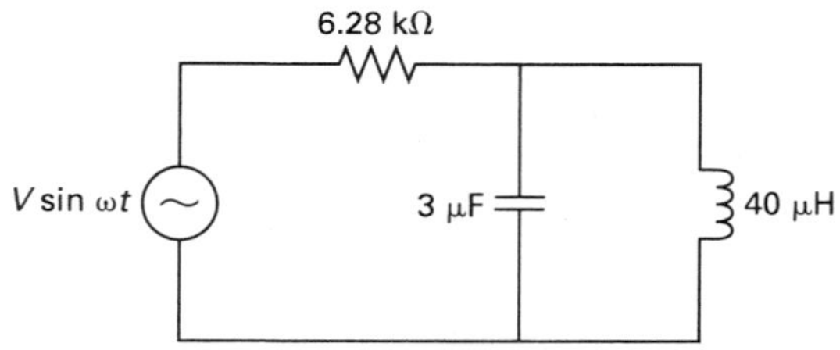

(A) 15 kHz
(B) 29 kHz
(C) 46 kHz
(D) 91 kHz

9. The switch in the series-RC circuit shown has been open for a long time, and there is no initial charge on the capacitor.

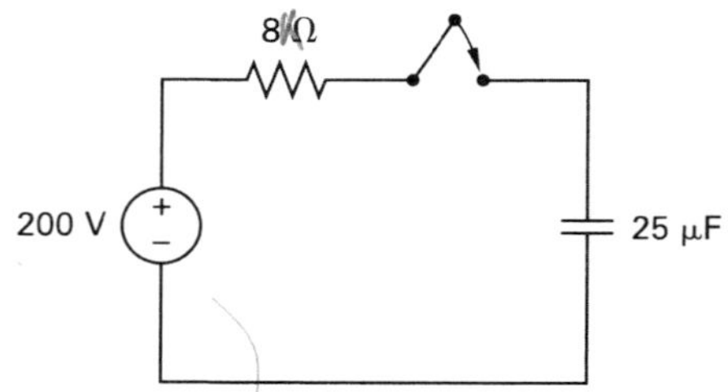

The switch is closed at time $t = 0$ s. Most nearly, what is the current at time $t = 0.4$ s?

(A) 1.0 mA
(B) 2.1 mA
(C) 2.4 mA
(D) 3.4 mA

10. The switch in the circuit shown has been open for a long time, and there is no initial current through the inductor.

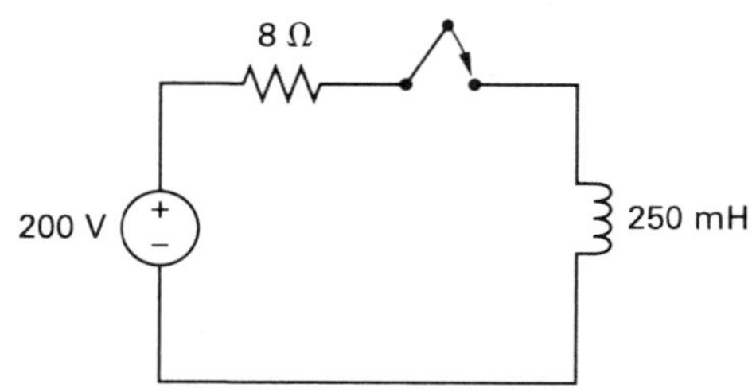

The switch is closed at time $t = 0$ s. Most nearly, what is the voltage across the inductor at time $t = 0.05$ s?

(A) 40 V

(B) 62 V

(C) 130 V

(D) 180 V

11. The circuit shown has a resonant frequency of 577 rad/s. The bandwidth is 6.66 rad/s.

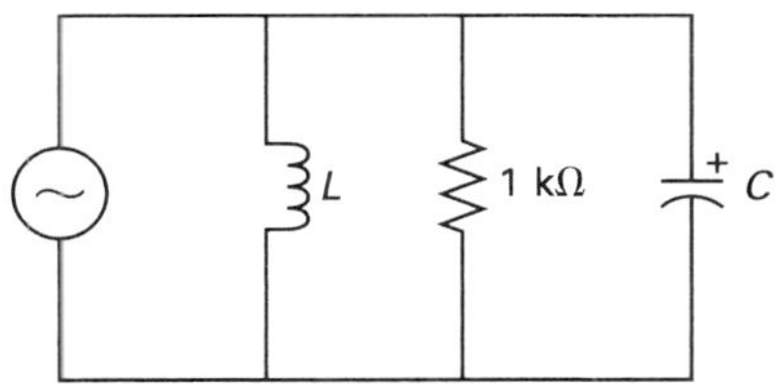

Most nearly, the capacitance is

(A) 150 μF

(B) 200 μF

(C) 250 μF

(D) 300 μF

12. In the series-RLC circuit shown, what is the transfer function in the Laplace domain, $V_2(s)/V_1(s)$?

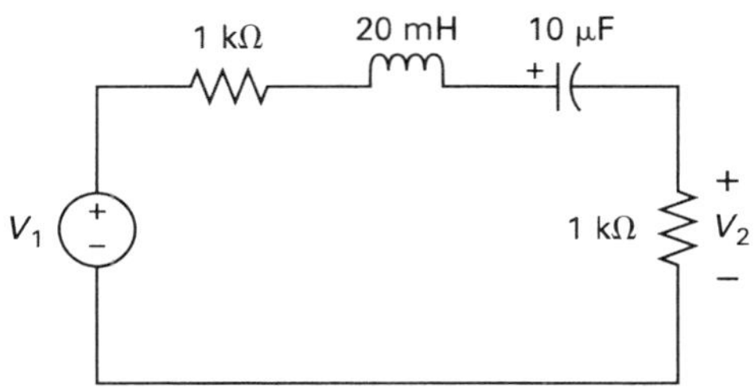

(A) $\dfrac{10^4 s}{s^2 + 10^5 s + 4 \times 10^6}$

(B) $\dfrac{4 \times 10^4 s}{s^2 + 10^5 s + 5 \times 10^6}$

(C) $\dfrac{5 \times 10^3 s}{s^2 + 2 \times 10^5 s + 5 \times 10^6}$

(D) $\dfrac{5 \times 10^4 s}{s^2 + 10^5 s + 5 \times 10^6}$

13. In the circuit shown, the capacitance, C, needed to achieve a power factor of 1.0 is most nearly

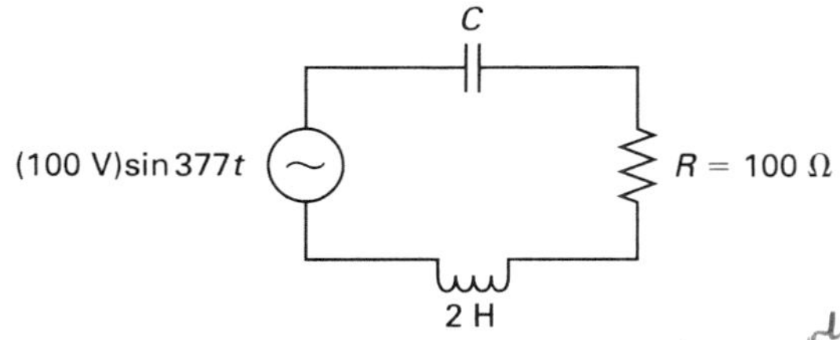

(A) 1.0 μF

(B) 1.6 μF

(C) 2.0 μF

(D) 3.5 μF

SOLUTIONS

1. Use the equation for transient inductor current. The exponent is

$$-\frac{Rt}{L} = -\frac{(2\times 10^3\ \Omega)(60\times 10^{-6}\ \text{s})}{80\times 10^{-3}\ \text{H}} = -1.5$$

The current passing through the conductor at $t = 60\ \mu\text{s}$ is

$$\begin{aligned} i_L(t) &= \left(\frac{V}{R}\right)(1-e^{-Rt/L}) \\ i_L(60\ \mu\text{s}) &= \left(\frac{10\ \text{V}}{2\times 10^3\ \Omega}\right)(1-e^{-1.5})\left(10^3\ \frac{\text{mA}}{\text{A}}\right) \\ &= 3.88\ \text{mA} \quad (3.9\ \text{mA}) \end{aligned}$$

The answer is (C).

2. R_1 and C_1 act as a low-pass filter with a cutoff frequency of 318 Hz. R_2 and C_2 act as a high-pass filter whose cutoff frequency is also 318 Hz. The isolating network prevents R_2 and C_2 from loading R_1 and C_1.

The combination acts as a band-pass filter because all frequencies significantly above and below 318 Hz are attenuated.

The answer is (D).

3. At the end of the transient period (after the switch is closed), the capacitor acts like an open circuit. There is no current through the open-circuit capacitor. Therefore, the voltage across the capacitor is the same as the voltage across the parallel 2 kΩ resistor.

Use the voltage-divider formula.

$$V_C = (40\ \text{V})\left(\frac{2\ \text{k}\Omega}{2\ \text{k}\Omega + 2\ \text{k}\Omega}\right) = 20\ \text{V}$$

The answer is (B).

4. The circuit is a low-pass filter with a cutoff frequency of

$$\frac{1}{RC} = \frac{1}{(50\ \Omega)(200\times 10^{-6}\ \text{F})} = 100\ \text{rad/s}$$

The input is at the cutoff frequency. The cutoff frequency is the half-power point, so the amplitude of the output voltage is the input voltage divided by $\sqrt{2}$.

$$v_{\text{out}} = \frac{v_{\text{in}}}{\sqrt{2}} = \frac{40\ \text{V}}{\sqrt{2}} = 28.3\ \text{V} \quad (28\ \text{V})$$

Alternatively, the voltage can be found from the voltage divider concept. The reactance of the capacitor is

$$\begin{aligned} X_C &= -\frac{1}{\omega C} = -\frac{1}{\left(100\ \frac{\text{rad}}{\text{s}}\right)(200\times 10^{-6}\ \text{F})} \\ &= -50\ \Omega \end{aligned}$$

The output amplitude is calculated using the voltage divider concept.

$$\begin{aligned} v_{\text{out}} &= v_{\text{in}}\left(\frac{|X_C|}{|R+jX_C|}\right) = v_{\text{in}}\left(\frac{|X_C|}{\sqrt{R^2+X_C^2}}\right) \\ &= (40\ \text{V})\left(\frac{|-50\ \Omega|}{\sqrt{(50\ \Omega)^2+(-50\ \Omega)^2}}\right) \\ &= 28.3\ \text{V} \quad (28\ \text{V}) \end{aligned}$$

The answer is (C).

5. Solve the decaying voltage equation for $v(t) = 2$ V$v(t) = 2$ V.

$$v(t) = 2\ \text{V} = 20e^{-t/400\times 10^{-3}\ \text{s}}$$

Substitute and divide both sides of the equation by 20 V.

$$\begin{aligned} \frac{2\ \text{V}}{20\ \text{V}} &= \frac{(20\ \text{V})e^{-t/400\times 10^{-3}\ \text{s}}}{20\ \text{V}} \\ 0.1 &= e^{-t/400\times 10^{-3}\ \text{s}} \end{aligned}$$

Take the natural logarithm of both sides of the equation.

$$\begin{aligned} \ln 0.1 &= \ln e^{-t/400\times 10^{-3}\ \text{s}} \\ &= -\frac{t}{400\times 10^{-3}\ \text{s}} \end{aligned}$$

Solve for t.

$$\begin{aligned} t &= -\left((400\times 10^{-3}\ \text{s})\left(1000\ \frac{\text{ms}}{\text{s}}\right)\right)\ln 0.1 \\ &= 921\ \text{ms} \quad (920\ \text{ms}) \end{aligned}$$

The answer is (C).

6. Since the circuit is initially open, the initial voltage on the capacitor is zero.

$$\begin{aligned} v_C(t) &= v_C(0)e^{-t/RC} + V(1 - e^{-t/RC}) \\ 0.8 &= 1 - e^{-t/(150\ \Omega)(100\times 10^{-6}\ \text{F})} \\ 0.2 &= e^{-t/0.015\ \text{s}} \end{aligned}$$

Take the natural log of both sides.

$$\begin{aligned} \ln 0.2 &= \frac{-t}{0.015\ \text{s}} \\ t &= 0.024\ \text{s} \quad (24\ \text{ms}) \end{aligned}$$

The answer is (D).

7. For a series-RLC circuit, the resonant frequency is

$$\begin{aligned} \omega_0 &= \frac{1}{\sqrt{LC}} = \frac{1}{\sqrt{(2\ \text{H})(1.5\times 10^{-6}\ \text{F})}} \\ &= 577\ \text{rad/s} \quad (580\ \text{rad/s}) \end{aligned}$$

The answer is (D).

8. The resonant frequency is

$$\begin{aligned} \omega_0 &= 2\pi f_0 = \frac{1}{\sqrt{LC}} \\ f_0 &= \frac{1}{2\pi\sqrt{LC}} \\ &= \frac{1}{2\pi\sqrt{(40\times 10^{-6}\ \text{H})(3\times 10^{-6}\ \text{F})}} \\ &= 14\,592\ \text{Hz} \quad (15\ \text{kHz}) \end{aligned}$$

The answer is (A).

9. Use the equation for transient current in an RC circuit. The exponent is

$$\begin{aligned} -t/RC &= \frac{-0.4\ \text{s}}{\left((8\ \text{k}\Omega)\left(1000\ \frac{\Omega}{\text{k}\Omega}\right)\right)\left(\dfrac{25\ \mu\text{F}}{10^6\ \frac{\text{F}}{\mu\text{F}}}\right)} \\ &= -2 \end{aligned}$$

The current at $t = 0.4$ s is

$$\begin{aligned} i(t) &= \{[V - v_C(0)]/R\}e^{-t/RC} \\ &= \left(\frac{200\ \text{V} - 0\ \text{V}}{(8\ \text{k}\Omega)\left(1000\ \frac{\Omega}{\text{k}\Omega}\right)}\right)e^{-2}\left(1000\ \frac{\text{mA}}{\text{A}}\right) \\ &= 3.38\ \text{mA} \quad (3.4\ \text{mA}) \end{aligned}$$

The answer is (D).

10. Use the equation for transient voltage in an RL circuit. The exponent is

$$\begin{aligned} -Rt/L &= \frac{-(8\ \Omega)(0.05\ \text{s})}{\dfrac{250\ \text{mH}}{1000\ \frac{\text{mH}}{\text{H}}}} \\ &= -1.6 \end{aligned}$$

The voltage on the inductor is

$$\begin{aligned} v_L(t) &= -i(0)Re^{-Rt/L} + Ve^{-Rt/L} \\ &= 0\ \text{V} + (200\ \text{V})e^{-1.6} \\ &= 40.38\ \text{V} \quad (40\ \text{V}) \end{aligned}$$

The answer is (A).

11. Rearrange the bandwidth equation to solve for the quality factor.

$$\begin{aligned} \text{BW} &= \frac{\omega_0}{Q} \\ Q &= \frac{\omega_0}{\text{BW}} = \frac{577\ \frac{\text{rad}}{\text{s}}}{6.66\ \frac{\text{rad}}{\text{s}}} \\ &= 86.64 \end{aligned}$$

Rearrange the equation for the quality factor in a parallel-RLC circuit to calculate the capacitance.

$$\begin{aligned} Q &= \omega_0 RC \\ C &= \frac{Q}{\omega_0 R} = \frac{86.64}{\left(577\ \frac{\text{rad}}{\text{s}}\right)(1\times 10^3\ \Omega)} \\ &= 150.15\times 10^{-6}\ \text{F} \quad (150\times \mu\text{F}) \end{aligned}$$

The answer is (A).

Circuit Analysis/ Linear Systems

12. This is a band-pass filter series circuit. The equivalent series resistance is

$$R_S = R_1 + R_2 = 1\ \text{k}\Omega + 1\ \text{k}\Omega = 2\ \text{k}\Omega$$

Use the equation for the transfer function of a band-pass filter series circuit.

$$\begin{aligned}\mathbf{H}(s) = \frac{\mathbf{V}_2(s)}{\mathbf{V}_1(s)} &= \frac{R_2}{L}\cdot\frac{s}{s^2 + sR_S/L + 1/LC}\\ &= \left(\frac{(1\ \text{k}\Omega)\left(1000\ \frac{\Omega}{\text{k}\Omega}\right)\left(1000\ \frac{\text{mH}}{\text{H}}\right)}{20\ \text{mH}}\right)\\ &\quad\times\left(\frac{s}{s^2 + \frac{s(2\ \text{k}\Omega)\left(1000\ \frac{\Omega}{\text{k}\Omega}\right)\left(1000\ \frac{\text{mH}}{\text{H}}\right)}{20\ \text{mH}} + \frac{\left(1000\ \frac{\text{mH}}{\text{H}}\right)\left(10^6\ \frac{\mu\text{F}}{\text{F}}\right)}{(20\ \text{mH})(10\ \mu\text{F})}}\right)\\ &= \frac{5\times 10^4 s}{s^2 + 10^5 s + 5\times 10^6}\end{aligned}$$

The answer is (D).

13. A circuit has a power factor of 1.0 when it is resonant. Use the equation for a circuit at resonant frequency. The input frequency for this circuit is 377 rad/s, so use this value as the resonant frequency and solve for the capacitance.

$$\omega_0 L = \frac{1}{\omega_0 C}$$

$$\begin{aligned}C = \frac{1}{\omega_0^2 L} &= \frac{1}{\left(377\ \frac{\text{rad}}{\text{s}}\right)^2 (2\ \text{H})}\\ &= 3.52\times 10^{-6}\ \text{F}\quad (3.5\ \mu\text{F})\end{aligned}$$

The answer is (D).

26 Three-Phase Power

PRACTICE PROBLEMS

1. The balanced delta load shown is connected to a stable, three-phase voltage source. The rms line voltage is 120 V.

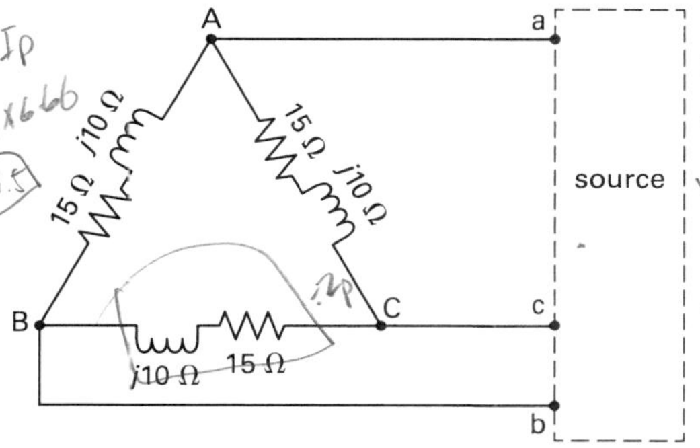

What is most nearly the rms magnitude of the line current?

(A) 4.80 A

(B) 8.50 A

(C) 11.5 A

(D) 12.0 A

2. A 240 V$\angle 0°$ source supplies a balanced, three-phase load in a delta configuration. The line current is 3 A$\angle 45°$. What is most nearly the magnitude of the complex power measured across the load?

(A) 420 VA

(B) 720 VA

(C) 1200 VA

(D) 2200 VA

3. A balanced, three-phase, delta-connected system has three identical phase impedances. The line-to-line voltage is 240 V. The line current is 41 A. Most nearly, the magnitudes of the phase voltage and phase current in the system are

(A) 138 V and 11.8 A

(B) 138 V and 23.7 A

(C) 240 V and 11.8 A

(D) 240 V and 23.7 A

4. A balanced, three-phase, wye-connected load system has three identical phase impedances. Most nearly, the equivalent delta-connected phase impedances are

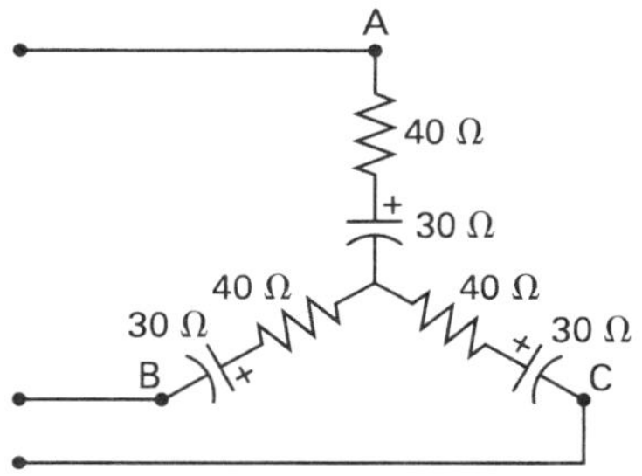

(A) 40 Ω − j30 Ω

(B) 40 Ω + j30 Ω

(C) 120 Ω − j90 Ω

(D) 120 Ω + j90 Ω

5. A balanced, three-phase, wye-connected system has three identical phase impedances. The line-to-line voltage is 240 V. The line-to-neutral current is 41 A. The power factor is 0.8 leading. Most nearly, the total apparent power in the system is

(A) 7.9 kW − j5.9 kVA

(B) 7.9 kW + j5.9 kVA

(C) 14 kW − j10 kVA

(D) 14 kW + j10 kVA

6. Three identical loads of $4\ \Omega - j3\ \Omega$ are delta-connected in the three-phase system. The line-to-line voltage is 120 V. Most nearly, the total apparent power in polar form is

(A) $2880\ \text{VA}\angle -36.87°$

(B) $2880\ \text{VA}\angle 36.87°$

(C) $8640\ \text{VA}\angle -36.87°$

(D) $8640\ \text{VA}\angle 36.87°$

7. The A phase of a balanced, three-phase, positive sequence system is taken as the reference. The line-to-neutral voltage is 277 V. The magnitude and phase angle of the CA line-to-line phase voltage is most nearly

(A) $277\ \text{V}\angle -90°$

(B) $277\ \text{V}\angle 150°$

(C) $480\ \text{V}\angle -90°$

(D) $480\ \text{V}\angle 150°$

SOLUTIONS

1. Calculate the magnitude of the impedance for the phase.

$$\begin{aligned}\mathbf{Z}_P &\equiv a + jb\\ |\mathbf{Z}_P| &= \sqrt{a^2 + b^2}\\ &= \sqrt{(15\ \Omega)^2 + (10\ \Omega)^2}\\ &= 18\ \Omega\end{aligned}$$

Find the magnitude of the phase current using Ohm's law for AC circuits.

$$\begin{aligned}\mathbf{V}_P &= \mathbf{I}_P\mathbf{Z}_P\\ |\mathbf{I}_P| &= \left|\frac{\mathbf{V}_P}{\mathbf{Z}_P}\right| = \frac{120\ \text{V}}{18\ \Omega}\\ &= 6.66\ \text{A}\end{aligned}$$

Solve for the magnitude of the line current using the delta line-phase relationship.

$$\begin{aligned}|\mathbf{I}_L| &= \sqrt{3}\,|\mathbf{I}_P| = \sqrt{3}\,(6.66\ \text{A})\\ &= 11.5\ \text{A}\end{aligned}$$

The answer is (C).

2. For a delta configuration, $V_L = V_P$. The magnitude of the complex power is

$$\begin{aligned}|\mathbf{S}| &= \sqrt{3}\,V_LI_L\\ &= \sqrt{3}\,(240\ \text{V})(3\ \text{A})\\ &= 1247\ \text{VA}\quad (1200\ \text{VA})\end{aligned}$$

The answer is (C).

3. The phase voltage is the line-to-line voltage.

$$V_L = V_P = 240\ \text{V}$$

The phase current is the line-to-line current.

$$\begin{aligned}I_L &= \sqrt{3}\,I_P\\ I_P &= \frac{I_L}{\sqrt{3}} = \frac{41\ \text{A}}{\sqrt{3}} = 23.7\ \text{A}\end{aligned}$$

The answer is (D).

4. The wye-connected phase impedances are

$$\mathbf{Z}_Y = 40\ \Omega - j30\ \Omega$$

Use the equation for the delta-wye impedance relationship. The equivalent delta-connected phase impedances are

$$\begin{aligned}\mathbf{Z}_{\Delta} &= 3\mathbf{Z}_{\mathrm{Y}} = (3)(40\ \Omega - j30\ \Omega)\\ &= 120\ \Omega - j90\ \Omega\end{aligned}$$

The answer is (C).

5. The power factor is leading, so the phase angle of the current is positive relative to the voltage. Accordingly, the power angle is negative.

$$\begin{aligned}pf &= \cos\theta_P = 0.8\\ \theta_P &= -\arccos 0.8\\ &= -36.87^\circ\end{aligned}$$

The total apparent power is

$$\begin{aligned}S &= 3\mathbf{V}_P\mathbf{I}_P^* = \sqrt{3}\ V_L I_L(\cos\theta_P + j\sin\theta_P)\\ &= \frac{\sqrt{3}\,(240\ \mathrm{V}\angle 0^\circ)(41\ \mathrm{A}\angle 36.87^\circ)^*}{1000\ \frac{\mathrm{W}}{\mathrm{kW}}}\\ &= \frac{\sqrt{3}\,(240\ \mathrm{V}\angle 0^\circ)(41\ \mathrm{A}\angle -36.87^\circ)}{1000\ \frac{\mathrm{W}}{\mathrm{kW}}}\\ &= \frac{\sqrt{3}\,(240\ \mathrm{V}\angle 0^\circ)(41\ \mathrm{A})(0.8 + j\sin -36.87^\circ)}{1000\ \frac{\mathrm{W}}{\mathrm{kW}}}\\ &= 13.63\ \mathrm{kW} - j10.2\ \mathrm{kVA}\quad (14\ \mathrm{kW} - j10\ \mathrm{kVA})\end{aligned}$$

The answer is (C).

6. The total apparent power is

$$\begin{aligned}\mathrm{S} &= 3\frac{V_L^2}{Z_\Delta^*} = \frac{(3)(120\ \mathrm{V})^2}{(4\ \Omega - j3\ \Omega)^*} = \frac{(3)(120\ \mathrm{V})^2}{4\ \Omega + j3\ \Omega}\\ &= \frac{(3)(120\ \mathrm{V})^2}{\sqrt{(4\ \Omega)^2 + (3\ \Omega)^2}\ \angle\arctan\frac{3\ \Omega}{4\ \Omega}}\\ &= \frac{(3)(120\ \mathrm{V})^2}{5\ \Omega\angle 36.87^\circ}\\ &= 8640\ \mathrm{VA}\angle -36.87^\circ\end{aligned}$$

The answer is (C).

7. In a positive sequence, the CA phase is 30° + 120° = 150° ahead of the A phase in a positive sequence, and the phase voltage is $\sqrt{3}$ times the line-to-neutral voltage. The CA line-to-line phase voltage is

$$\begin{aligned}\mathbf{V}_{\mathrm{ca}} &= \sqrt{3}\ V_P\angle 150^\circ\\ &= \sqrt{3}\,(277\ \mathrm{V})\angle 150^\circ\\ &= 480\ \mathrm{V}\angle 150^\circ\end{aligned}$$

The answer is (D).

$$.5 = \frac{Z_1 - Z_0}{Z_1 + Z_0}$$

$$.5(Z_1 + Z_0) = Z_1 - Z_0$$

27 Transmission Lines

PRACTICE PROBLEMS

1. A transmission line has an inductance and capacitance of 2.5×10^{-6} H/m and 1.0×10^{-9} F/m, respectively. The line is terminated with a purely resistive load. The standing wave ratio at the load is 3.0. Most nearly, what is the load resistance?

(A) 25 Ω

(B) 50 Ω

(C) 100 Ω

(D) 150 Ω

2. A transmission line has a characteristic impedance of 75 Ω. The load at the end of the transmission line has a characteristic impedance of 377 Ω. Most nearly, the percentage of the signal voltage that is transmitted to the load is

(A) 10%

(B) 30%

(C) 50%

(D) 100%

3. A lossless transmission line has a characteristic impedance of 100 Ω for a 100 MHz wave traveling at 70% of the speed of light. Most nearly, the capacitance and inductance of the line per unit length are

(A) 29 pF/m and 0.29 μH/m

(B) 48 pF/m and 0.48 μH/m

(C) 29 pF/m and 0.96 μH/m

(D) 13 pF/m and 0.26 μH/m

4. A lossless transmission line has a propagation constant of 2.5 rad/m when transmitting a 50 MHz signal. Most nearly, what antenna length would be equivalent to one-fourth of a wavelength of the transmitted signal?

(A) 0.2π m

(B) 0.5π m

(C) 0.8π m

(D) 2.0π m

5. Which of these statements concerning transmission lines is FALSE?

(A) A standing wave is the sum of the incident and reflected waves.

(B) The reflection coefficient is the ratio of the reflected wave voltage to the incident wave voltage.

(C) The magnitude of a standing wave is constant at a given position of the transmission line.

(D) The standing wave ratio is the ratio of the maximum and minimum current magnitudes of a standing wave.

6. A sinusoidal voltage wave with an amplitude of 5 V propagates in a transmission line. The line has a characteristic impedance of 25 Ω. The transmission line is loaded at its end with an impedance of 50 Ω + j25 Ω. Most nearly, the voltage magnitude of the wave reflected from the load is

(A) 0.22 V

(B) 1.1 V

(C) 2.2 V

(D) 5.0 V

7. When is maximum power transferred from a generator to the load through a transmission line?

(A) when the load impedance is equal to the characteristic impedance of the line

(B) when the reflection coefficient is one

(C) when the standing wave ratio is zero

(D) all of the above

8. What are the reflection coefficients for transmission lines that have terminating loads of an open circuit and a short circuit?

(A) $\Gamma_{\text{open circuit}} = 0,\ \Gamma_{\text{short circuit}} = 1$

(B) $\Gamma_{\text{open circuit}} = 1,\ \Gamma_{\text{short circuit}} = -1$

(C) $\Gamma_{\text{open circuit}} = 1,\ \Gamma_{\text{short circuit}} = 0$

(D) $\Gamma_{\text{open circuit}} = 1,\ \Gamma_{\text{short circuit}} = 1$

9. A lossless transmission line has a length of 30 m and a characteristic impedance of 50 Ω. The line operates at 5 MHz with a load impedance of $60\ \Omega + j40\ \Omega$. The propagation constant is 1.396 rad/m. Most nearly, the input impedance seen by the source is

(A) $55.9\ \Omega\angle -31.0°$

(B) $55.9\ \Omega\angle -35.7°$

(C) $66.9\ \Omega\angle -35.7°$

(D) $69.9\ \Omega\angle -33.2°$

10. A lossless coaxial cable terminates with an open circuit load. What is the transmission line voltage expressed in terms of the voltage at the generator, V_{gen}, the transmission line length, d, and the propagation constant, β?

(A) $\dfrac{2V_{gen}}{\cos \beta d}$

(B) $\dfrac{2V_{gen}}{2d\cos \beta d}$

(C) $\dfrac{V_{gen}}{\cos \beta d}$

(D) $\dfrac{2\pi V_{gen}}{\cos \beta d}$

11. The equation for transmission line voltage is

$$\mathbf{V}(d) = \mathbf{V}^{+}e^{j\beta d} + \mathbf{V}^{-}e^{-j\beta d}$$

In this equation, what units are associated with the product of the propagation constant, β, and the transmission line length, d?

(A) 1/m

(B) m

(C) m/s

(D) none (dimensionless)

12. What are the units of the transmission line term?

(A) 1/Ω

(B) Ω

(C) Ω·m

(D) none (dimensionless)

SOLUTIONS

1. Use the standing wave ratio (SWR) equation to find the reflection coefficient, Γ, in terms of the SWR.

$$\text{SWR} = \frac{1+|\Gamma|}{1-|\Gamma|}$$

$$\Gamma = \frac{\text{SWR}-1}{\text{SWR}+1} = \frac{3.0-1}{3.0+1} = 0.5$$

Use the equation for the reflection coefficient.

$$\Gamma = \frac{Z_L - Z_0}{Z_L + Z_0}$$

$$\Gamma(Z_L + Z_0) = Z_L - Z_0$$

$$Z_0 + \Gamma Z_0 = Z_L - \Gamma Z_L$$

$$Z_0(1+\Gamma) = Z_L(1-\Gamma)$$

$$Z_L = Z_0\left(\frac{1+\Gamma}{1-\Gamma}\right) = \sqrt{\frac{L}{C}}\left(\frac{1+\Gamma}{1-\Gamma}\right) = \sqrt{\frac{2.5\times 10^{-6}\ \frac{\text{H}}{\text{m}}}{1.0\times 10^{-9}\ \frac{\text{F}}{\text{m}}}}\left(\frac{1+0.5}{1-0.5}\right) = 150\ \Omega$$

The answer is (D).

2. The reflection coefficient is

$$\Gamma = \left|\frac{Z_L - Z_0}{Z_L + Z_0}\right| = \left|\frac{377\ \Omega - 75\ \Omega}{377\ \Omega + 75\ \Omega}\right| = 0.668$$

The fraction of the voltage transmitted is one minus the reflection coefficient.

$$\%_{\text{transmitted}} = (1-\Gamma)\times 100\% = (1-0.668)\times 100\% = 33.2\%\quad (30\%)$$

The answer is (B).

3. For a lossless transmission line, the characteristic impedance is

$$\mathbf{Z}_0 = \sqrt{L/C}$$

Rearrange to solve for the capacitance.

$$C = \frac{L}{Z_0^2}$$

For a lossless transmission line, the velocity of propagation is

$$U = \frac{1}{\sqrt{LC}} = 0.7c$$
$$\sqrt{LC} = \frac{1}{0.7c}$$
$$LC = \frac{1}{(0.7c)^2}$$

Substitute capacitance from the characteristic impedance equation, and solve for the inductance.

$$L\left(\frac{L}{Z_0^2}\right) = \frac{1}{(0.7c)^2}$$
$$\frac{L}{Z_0} = \frac{1}{0.7c}$$
$$L = \frac{Z_0}{0.7c} = \frac{100\ \Omega}{(0.7)\left(3.0\times 10^8\ \frac{\text{m}}{\text{s}}\right)}$$
$$= 4.76\times 10^{-7}\ \text{H/m}\quad (0.48\ \mu\text{H/m})$$

The capacitance is

$$C = \frac{L}{Z_0^2} = \frac{4.76\times 10^{-7}\ \frac{\text{H}}{\text{m}}}{(100\ \Omega)^2}$$
$$= 4.76\times 10^{-11}\ \text{F/m}\quad (48\ \text{pF/m})$$

The answer is (B).

4. Use the equation for the propagation constant to calculate the wavelength of a lossless transmission line.

$$\beta = \frac{2\pi}{\lambda}$$
$$\lambda = \frac{2\pi}{\beta} = \frac{2\pi\ \text{rad}}{2.5\ \frac{\text{rad}}{\text{m}}}$$
$$= 0.8\pi\ \text{m}$$

One-fourth of a wavelength is

$$\frac{\lambda}{4} = \frac{0.8\pi\ \text{m}}{4} = 0.2\pi\ \text{m}$$

The answer is (A).

5. The magnitude of a standing wave at a given position along the transmission line will vary with time, so option C is false. The locations of the maxima and minima in a standing wave are constant, but the overall magnitude of the wave changes.

A standing wave is the sum of the incident and reflected waves, so option A is true.

The ratio of the voltages of the reflected and incident waves is the reflection coefficient, so option B is true.

The standing wave ratio is the ratio of the maximum magnitude to the minimum magnitude of a standing wave for both voltage and current, so option D is true.

The answer is (C).

6. The reflection coefficient is the ratio of the reflected wave voltage to the incident wave voltage.

$$\Gamma = \frac{\mathbf{V}^-}{\mathbf{V}^+} = \frac{\mathbf{Z}_L - \mathbf{Z}_0}{\mathbf{Z}_L + \mathbf{Z}_0}$$
$$\mathbf{V}^- = \mathbf{V}^+\left(\frac{\mathbf{Z}_L - \mathbf{Z}_0}{\mathbf{Z}_L + \mathbf{Z}_0}\right)$$
$$= (5\ \text{V})\left(\frac{50\ \Omega + j25\ \Omega - 25\ \Omega}{50\ \Omega + j25\ \Omega + 25\ \Omega}\right)$$
$$= (5\ \text{V})\left(\frac{1+j}{3+j}\right)\left(\frac{3-j}{3-j}\right)$$
$$= (5\ \text{V})\left(\frac{4+2j}{10}\right)$$
$$= (2+j)\ \text{V}$$
$$|V^-| = \sqrt{(2\ \text{V})^2 + (1\ \text{V})^2} = 2.2\ \text{V}$$

The answer is (C).

7. When maximum power is transferred to the load, the incident wave is entirely absorbed by the load and there is no reflected power. This happens only when the characteristic impedance of the line, $\mathbf{Z}_0$, and the load impedance, $\mathbf{Z}_L$, are the same. Option A is true.

The reflection coefficient, Γ, is equal to $(\mathbf{Z}_L - \mathbf{Z}_0)/(\mathbf{Z}_L + \mathbf{Z}_0)$. When $\mathbf{Z}_L$ and $\mathbf{Z}_0$ are equal, the reflection coefficient is zero. Option B is false.

The standing wave ratio is equal to $(1+|\Gamma|)/(1-|\Gamma|)$. When the reflection coefficient is zero, the standing wave ratio is one. Option C is false.

The answer is (A).

8. When the load is an open circuit, the load impedance, $\mathbf{Z}_L$, is infinite, so the reflection coefficient is

$$\Gamma_{\text{open circuit}} = \lim_{\mathbf{Z}_L\to\infty} \frac{\mathbf{Z}_L - \mathbf{Z}_0}{\mathbf{Z}_L + \mathbf{Z}_0} = 1$$

Power

When the load is a short circuit, the load impedance, $\mathbf{Z}_L$, is zero, so the reflection coefficient is

$$\Gamma_{\text{short circuit}} = \frac{\mathbf{Z}_L - \mathbf{Z}_0}{\mathbf{Z}_L + \mathbf{Z}_0} = \frac{0 - \mathbf{Z}_0}{0 + \mathbf{Z}_0} = -1$$

The answer is (B).

9. Use the equation for transmission line impedance at a point. The phase angle of the impedance at the source is

$$\beta d = \left(\left(1.396\ \frac{\text{rad}}{\text{m}}\right)\left(\frac{180°}{\pi}\right)\right)(30\ \text{m}) = 2400°$$

The input impedance seen by the source is

$$\begin{aligned}\mathbf{Z}_{\text{in}}(d) &= \mathbf{Z}_0\frac{\mathbf{Z}_L + j\mathbf{Z}_0\tan(\beta d)}{\mathbf{Z}_0 + j\mathbf{Z}_L\tan(\beta d)}\\ &= (50\ \Omega)\left(\frac{60\ \Omega + j40\ \Omega + (j50\ \Omega)\tan 2400°}{50\ \Omega + (j60\ \Omega - 40\ \Omega)\tan 2400°}\right)\\ &= (50\ \Omega)\left(\frac{60 + j125}{-18 + j102}\right)\left(\frac{-18 - j102}{-18 - j102}\right)\\ &= (50\ \Omega)\left(\frac{11{,}652 + j8370}{10{,}728}\right)\\ &= 54.3\ \Omega + j39.01\ \Omega\end{aligned}$$

Converting from rectangular form to polar form,

$$\begin{aligned}r &= \sqrt{(54.3\ \Omega)^2 + (39.01\ \Omega)^2} = 66.86\ \Omega\\ \theta &= \arctan\frac{39.01\ \Omega}{54.3\ \Omega} = 35.69°\\ Z_{\text{in}}(d) &= 66.86\ \Omega\angle{-35.69°}\quad (66.9\ \Omega\angle{-35.7°})\end{aligned}$$

The answer is (C).

10. The voltage equation for a lossless transmission line is

$$\mathbf{V}(d) = \mathbf{V}^+e^{j\beta d} + \mathbf{V}^-e^{-j\beta d}$$

Let $d = 0$ at the termination. Let $V_1 = \mathbf{V}^+(0)$ and $V_2 = \mathbf{V}^-(0)$. The load is an open circuit, so the reflection coefficient is 1.

$$\begin{aligned}\Gamma &= \frac{\mathbf{V}^-}{\mathbf{V}^+} = \frac{V_1}{V_2} = 1\\ V_1 &= V_2\end{aligned}$$

The voltage equation for this lossless transmission line is

$$\begin{aligned}\mathbf{V}(d) &= V_1e^{j\beta d} + V_1e^{-j\beta d}\\ &= V_1(e^{j\beta d} + e^{-j\beta d})\end{aligned}$$

Apply Euler's identity to the part in the parentheses.

$$\mathbf{V}(d) = V_1(2\cos\beta d)$$

The voltage at the generator is V_{gen}, so

$$\begin{aligned}\mathbf{V}(d) &= V_{\text{gen}} = V_1(2\cos\beta d)\\ V_1 &= \frac{V_{\text{gen}}}{2\cos\beta d}\end{aligned}$$

The voltage at the load end is

$$\begin{aligned}\mathbf{V}(0) &= V_1(2\cos 0) = 2V_1\\ &= 2\left(\frac{V_{\text{gen}}}{2\cos\beta d}\right)\\ &= \frac{V_{\text{gen}}}{\cos\beta d}\end{aligned}$$

The answer is (C).

11. The units for the propagation constant, β, are the reciprocal of meters, 1/m or m^{-1}. The units for the distance, d, are meters, m. The product βd is dimensionless. (Any term used as the argument for a trigonometric function or as an exponent must be dimensionless.)

The answer is (D).

12. The transmission line term is an impedance term, so the units are ohms (Ω).

The answer is (B).

Power

28 Power Distribution and Overcurrent Protection

PRACTICE PROBLEMS

1. The point of common coupling—also known as the service connection point—between a utility electric distribution system and a customer's electric distribution system is most commonly

(A) a transformer

(B) an electric meter

(C) a fused disconnect switch or circuit breaker

(D) the main overcurrent protective device

2. A step-up transformer is used to connect a large generating plant to the electric grid in order to

(A) prevent ground faults at the generator terminal

(B) provide a means of disconnecting the generator from the grid

(C) maintain stability

(D) increase the output voltage of the generator to match the transmission voltage at the interconnection point

3. Which of these statements about a simple radial distribution system is FALSE?

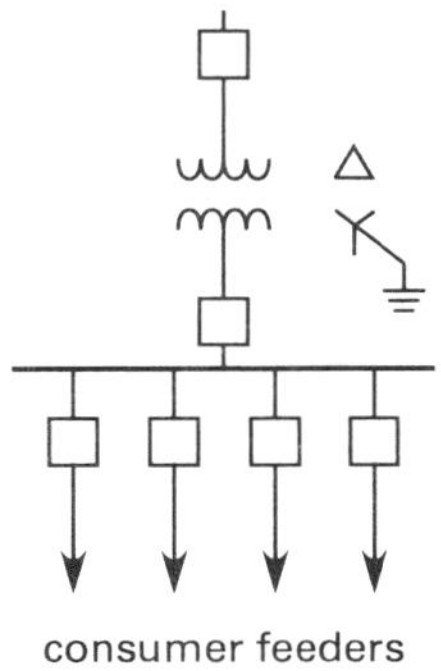

(A) A single primary service and transformer supply all consumer feeders.

(B) There is no redundancy of equipment (i.e., switches, fuses, transformers, service cable).

(C) Operation and expansion are simpler than in a looped or networked system.

(D) A backup generator must be connected on the primary side of the supply transformer.

4. When deciding on whether to use a fuse or a circuit breaker for overcurrent protection, which factor should generally NOT be considered?

(A) environmental regulations

(B) interrupting ratings

(C) selective coordination

(D) economics

SOLUTIONS

1. The point of common coupling is defined by the *National Electrical Code* and local regulations. It is the point where the jurisdiction of the local utility ends and that of the local electrical authority (the customer) begins. The point of common coupling is most commonly the site of an electric meter owned by the utility.

A transformer may be located either upstream of the meter (in which case it is owned by the utility) or downstream of the meter (in which case it is owned by the customer), so option A is incorrect.

A fused disconnect switch, circuit breaker, or main overcurrent protective device is owned by the customer and will be downstream of the meter, so options C and D are incorrect.

The answer is (B).

2. The output voltage of a large generator is in the range of 3 kV to 13.8 kV. The output voltage must be increased for interconnection with the electric grid, usually at transmission-level voltages in the range of 138 kV to 765 kV. The step-up transformer provides this voltage transformation. Option D is correct.

The step-up transformer does not prevent ground faults at the generator terminal. Ground faults at the generator terminal are minimized by impedance grounding. Option A is incorrect.

The step-up transformer is not a means of disconnection. This function is provided by breakers or switches on the high and low sides of the transformer. Option B is incorrect.

Stability is maintained by the generator's control system, not by the step-up transformer. Option C is incorrect.

The answer is (D).

3. There are several types of industrial and commercial electric distribution systems in use. The simplest is a radial distribution system, in which loads are supplied through a single electrical path. In the system shown, the loads are supplied through one transformer and one common bus. Each feeder has a means of disconnect and overcurrent protection as well as upstream breakers that can operate to isolate the transformer and the common bus. Option A is true.

From a load perspective, there is no redundancy of equipment. A fault on any device upstream of the load will cause an outage. Option B is true.

Operation and expansion are much simpler than in looped and networked distribution systems. Option C is true.

With the proper controls and protection devices, backup generation can be connected on any of the consumer feeders. Option D is false.

The answer is (D).

4. Fuses and circuit breakers are equally valid choices for overcurrent protection. There is no specific rule for determining whether to use one or the other. However, depending on the particular situation, one may be a better choice over the other.

Both fuses and circuit breakers meet environmental regulations. When deciding between fuses and circuit breakers, such regulations are not a consideration. Option A is correct.

The interrupting rating and the need and ability for selective coordination should be considered. Options B and C are incorrect.

Once it has been determined that either fuses or circuit breakers are acceptable for the application, the economics of a project should be taken into consideration. Option D is incorrect.

The answer is (A).

29 Motors and Generators

PRACTICE PROBLEMS

1. A motor with three pole pairs is operating at 60 Hz. The rotational speed is 1080 rpm. Most nearly, the slip angular velocity is

(A) π rad/s

(B) $2\pi/3$ rad/s

(C) 2π rad/s

(D) 4π rad/s

2. A three-phase synchronous generator produces a 60 Hz waveform when operating at 1200 rpm. How many poles does the generator have?

(A) 3

(B) 4

(C) 6

(D) 12

3. A synchronous motor is running at 1200 rpm on a 60 Hz supply voltage. The supply frequency is changed to 50 Hz. Most nearly, the mechanical angular velocity when the motor reaches steady state is

(A) 18 rad/s

(B) 65 rad/s

(C) 100 rad/s

(D) 380 rad/s

4. A DC generator turns at 2000 rpm and has an output of 200 V. The armature constant is 0.5 V·min/Wb, and the field constant of the machine is 0.02 H. Most nearly, what is the field current produced by the generator?

(A) 4.0 A

(B) 10 A

(C) 1000 A

(D) 4000 A

5. A two-pole induction motor operates on a three-phase, 60 Hz line-to-line supply. The motor speed is 3420 rpm, and the rms voltage is 240 V. Most nearly, what is the slip?

(A) 5%

(B) 7%

(C) 10%

(D) 20%

6. A DC generator produces 24 V while operating at 1200 rpm with a magnetic flux of 0.02 Wb. The same generator is operated at 1000 rpm with a magnetic flux of 0.05 Wb. Disregarding armature resistance, most nearly what is the new voltage the generator produces?

(A) 20 V

(B) 50 V

(C) 60 V

(D) 100 V

7. If the slip in a motor decreases while the torque remains constant, what is the effect on the motor's speed and power?

(A) speed increases, power increases

(B) speed increases, power decreases

(C) speed decreases, power increases

(D) speed decreases, power decreases

8. A four-pole induction motor operates on a three-phase, 60 Hz line-to-line supply. The rms voltage is 240 V. The slip is 2%. What is most nearly the operating speed?

(A) 1240 rpm

(B) 1660 rpm

(C) 1760 rpm

(D) 1800 rpm

9. A four-pole AC motor operates at a synchronous speed of 1800 rpm when running on a three-phase, 240 V_{rms} line-to-line supply. Most nearly, the electrical frequency of the supply voltage is

(A) 25 Hz

(B) 30 Hz

(C) 50 Hz

(D) 60 Hz

10. A six-pole induction motor operates on a three-phase, 60 Hz, 120 V_{rms} line-to-line supply. The percent slip is 4%. Most nearly, the speed of operation is

(A) 1100 rpm

(B) 1150 rpm

(C) 1180 rpm

(D) 1200 rpm

11. A DC generator operates at 1000 rpm with a magnetic flux of 0.03 Wb at the poles. The armature constant of the generator is 1 V·min/Wb. Most nearly, the voltage generated is

(A) 3.0 V

(B) 30 V

(C) 48 V

(D) 60 V

12. A 10 hp DC motor draws 42 A. Most nearly, what is the armature voltage?

(A) 42 V

(B) 89 V

(C) 120 V

(D) 180 V

13. The armature of a DC motor draws 50 A while generating a magnetic flux of 0.03 Wb. The armature constant of the motor is 1 V·min/Wb. Most nearly, the torque developed is

(A) 10 N·m

(B) 12 N·m

(C) 14 N·m

(D) 19 N·m

14. A DC servomotor rotates at 50 rpm when drawing 4 A from an input voltage of 30 V. The winding resistance is 2 Ω. Most nearly, what torque is developed by the motor?

(A) 1.8 N·m

(B) 8.4 N·m

(C) 12 N·m

(D) 17 N·m

SOLUTIONS

1. There are three pole pairs, so the number of poles is $p = 6$. The synchronous speed is

$$\begin{aligned} n_s &= 120f/p \\ &= \frac{\left(120\ \frac{\frac{\text{rev}}{\text{min}}}{\text{Hz}}\right)(60\ \text{Hz})}{6} \\ &= 1200\ \text{rev/min} \end{aligned}$$

The fractional slip is

$$\begin{aligned} \text{slip} &= (n_s - n)/n_s \\ &= \frac{1200\ \frac{\text{rev}}{\text{min}} - 1080\ \frac{\text{rev}}{\text{min}}}{1200\ \frac{\text{rev}}{\text{min}}} \\ &= 0.1 \end{aligned}$$

The angular velocity of the slip is

$$2\pi(\text{slip})n_s = \frac{\left(2\pi\ \frac{\text{rad}}{\text{rev}}\right)(0.1)\left(1200\ \frac{\text{rev}}{\text{min}}\right)}{60\ \frac{\text{s}}{\text{min}}} = 4\pi\ \text{rad/s}$$

The answer is (D).

2. Use the equation for synchronous speed, and solve for the number of poles.

$$n_s = 120f/p$$

$$p = \frac{120f}{n_s} = \frac{\left(120\ \frac{\frac{\text{rev}}{\text{min}}}{\text{Hz}}\right)(60\ \text{Hz})}{1200\ \frac{\text{rev}}{\text{min}}} = 6$$

The answer is (C).

3. When the motor reaches steady state, it is turning at the synchronous speed. The synchronous speed is related linearly to the supply frequency.

$$\frac{n_{s2}}{n_{s1}} = \frac{f_2}{f_1}$$

$$\begin{aligned} n_{s2} &= \left(\frac{f_2}{f_1}\right)n_{s1} = \left(\frac{50\ \text{Hz}}{60\ \text{Hz}}\right)\left(1200\ \frac{\text{rev}}{\text{min}}\right) \\ &= 1000\ \text{rev/min} \end{aligned}$$

Convert the synchronous speed in revolutions per minute to angular velocity in radians per second.

$$\begin{aligned} \Omega_2 &= \frac{n_{s2}\left(2\pi\ \frac{\text{rad}}{\text{rev}}\right)}{60\ \frac{\text{s}}{\text{min}}} = \frac{\left(1000\ \frac{\text{rev}}{\text{min}}\right)\left(2\pi\ \frac{\text{rad}}{\text{rev}}\right)}{60\ \frac{\text{s}}{\text{min}}} \\ &= 104.7\ \text{rad/s} \quad (100\ \text{rad/s}) \end{aligned}$$

The answer is (C).

4. The magnetic flux is

$$\begin{aligned} V_a &= K_a n\phi \\ \phi &= \frac{V_a}{K_a n} = \frac{200\ \text{V}}{\left(0.5\ \frac{\text{V·min}}{\text{Wb}}\right)\left(2000\ \frac{\text{rev}}{\text{min}}\right)} \\ &= 0.2\ \text{Wb} \end{aligned}$$

Solve for the field current.

$$\begin{aligned} \phi &= K_f I_f \\ I_f &= \frac{\phi}{K_f} = \frac{0.2\ \text{Wb}}{0.02\ \text{H}} \\ &= 10\ \text{A} \end{aligned}$$

The answer is (B).

5. The synchronous speed is

$$\begin{aligned} n_s &= 120f/p \\ &= \frac{\left(120\ \frac{\frac{\text{rev}}{\text{min}}}{\text{Hz}}\right)(60\ \text{Hz})}{2} \\ &= 3600\ \text{rev/min} \end{aligned}$$

The slip is

$$\begin{aligned} \text{slip} &= (n_s - n)/n_s \times 100\% \\ &= \frac{3600\ \frac{\text{rev}}{\text{min}} - 3420\ \frac{\text{rev}}{\text{min}}}{3600\ \frac{\text{rev}}{\text{min}}} \times 100\% \\ &= 5\% \end{aligned}$$

The answer is (A).

Power

6. Since the generated voltage is $V_a = K_a n\phi$, the voltage is proportional to the rotational speed and the flux. The new generated voltage will be

$$\begin{aligned} V_{\text{new}} &= V_{\text{old}}\left(\frac{n_{\text{new}}}{n_{\text{old}}}\right)\left(\frac{\phi_{\text{new}}}{\phi_{\text{old}}}\right) \\ &= (24\ \text{V})\left(\frac{1000\ \frac{\text{rev}}{\text{min}}}{1200\ \frac{\text{rev}}{\text{min}}}\right)\left(\frac{0.05\ \text{Wb}}{0.02\ \text{Wb}}\right) \\ &= 50\ \text{V} \end{aligned}$$

The answer is (B).

7. Solving the slip equation for the motor speed yields

$$\begin{aligned} \text{slip} &= (n_s - n)/n_s \\ n &= n_s(1 - \text{slip}) \end{aligned}$$

Therefore, the speed of the motor increases as the slip decreases.

Find the relationship between slip and power, P.

$$\begin{aligned} P = T\Omega &= T\left(\frac{60\ \frac{\text{s}}{\text{min}}}{2\pi\ \frac{\text{rad}}{\text{rev}}}\right)n \\ &= T\left(\frac{60\ \frac{\text{s}}{\text{min}}}{2\pi\ \frac{\text{rad}}{\text{rev}}}\right)n_s(1 - \text{slip}) \end{aligned}$$

Therefore, the power of the motor increases as the slip decreases.

The answer is (A).

8. The synchronous speed is

$$\begin{aligned} n_s &= 120f/p \\ &= \frac{\left(120\ \frac{\frac{\text{rev}}{\text{min}}}{\text{Hz}}\right)(60\ \text{Hz})}{4} \\ &= 1800\ \text{rev/min} \end{aligned}$$

Solving the slip equation for the operating speed yields

$$\begin{aligned} \text{slip} &= (n_s - n)/n_s \\ n &= n_s(1 - \text{slip}) \\ &= \left(1800\ \frac{\text{rev}}{\text{min}}\right)(1 - 0.02) \\ &= 1764\ \text{rev/min} \quad (1760\ \text{rpm}) \end{aligned}$$

The answer is (C).

9. Use the equation for synchronous speed, and solve for the electrical frequency.

$$\begin{aligned} n_s &= 120f/p \\ f &= \frac{n_s p}{120} = \frac{\left(1800\ \frac{\text{rev}}{\text{min}}\right)(4)}{120\ \frac{\frac{\text{rev}}{\text{min}}}{\text{Hz}}} = 60\ \text{Hz} \end{aligned}$$

The answer is (D).

10. The synchronous speed is

$$\begin{aligned} n_s &= 120f/p \\ &= \frac{\left(120\ \frac{\frac{\text{rev}}{\text{min}}}{\text{Hz}}\right)(60\ \text{Hz})}{6} \\ &= 1200\ \text{rev/min} \end{aligned}$$

Use the equation for slip, and solve for the operating speed.

$$\begin{aligned} \text{slip} &= (n_s - n)/n_s \\ n &= n_s - (\text{slip})n_s = 1200\ \frac{\text{rev}}{\text{min}} - (0.04)\left(1200\ \frac{\text{rev}}{\text{min}}\right) \\ &= 1152\ \text{rev/min} \quad (1150\ \text{rpm}) \end{aligned}$$

The answer is (B).

11. Use the equation for armature voltage. The generated voltage is

$$\begin{aligned} V_a = K_a n\phi &= \left(1\ \frac{\text{V}\cdot\text{min}}{\text{Wb}}\right)\left(1000\ \frac{\text{rev}}{\text{min}}\right)(0.03\ \text{Wb}) \\ &= 30\ \text{V} \end{aligned}$$

The answer is (B).

Power

12. Use the equation for the mechanical power of a DC motor, and solve for the armature voltage.

$$P_m = V_a I_a$$

$$V_a = \frac{P_m}{I_a} = \frac{(10\ \text{hp})\left(745.7\ \frac{\text{W}}{\text{hp}}\right)}{42\ \text{A}}$$
$$= 177.5\ \text{V} \quad (180\ \text{V})$$

The answer is (D).

13. The torque is

$$T_m = (60/2\pi)K_a \phi I_a$$
$$= \left(\frac{60}{2\pi}\right)\left(1\ \frac{\text{V·min}}{\text{Wb}}\right)(0.03\ \text{Wb})(50\ \text{A})$$
$$= 14.32\ \text{N·m} \quad (14\ \text{N·m})$$

The answer is (C).

14. Use the equation for the voltage of a servomotor, and solve for the motor voltage constant.

$$V = IR + K_E\omega$$
$$K_E = \frac{V - IR}{\omega}$$
$$= \frac{(30\ \text{V} - (4\ \text{A})(2\ \Omega))\left(60\ \frac{\text{s}}{\text{min}}\right)}{\left(50\ \frac{\text{rev}}{\text{min}}\right)\left(2\pi\ \frac{\text{rad}}{\text{rev}}\right)}$$
$$= 4.2\ \text{V·s/rad}$$

The motor torque constant is equal to the motor electrical constant but is expressed in different units.

$$K_T = K_E = 4.2\ \text{V·s/rad} = 4.2\ \text{N·m/A}$$

The torque is

$$T = K_T I = \left(4.2\ \frac{\text{N·m}}{\text{A}}\right)(4\ \text{A})$$
$$= 16.8\ \text{N·m} \quad (17\ \text{N·m})$$

The answer is (D).

30 Semiconductor Devices and Circuits

PRACTICE PROBLEMS

1. At an equilibrium temperature of 50°C, the contact potential of a pn junction is 0.6 V. The intrinsic carrier concentration is $n_i = 1.5 \times 10^{10}\ 1/\text{cm}^3$, and the donor concentration in the n-type material is $7.0 \times 10^{17}\ 1/\text{cm}^3$. The acceptor concentration is most nearly

(A) $7.0 \times 10^{10}\ 1/\text{cm}^3$

(B) $1.0 \times 10^{11}\ 1/\text{cm}^3$

(C) $7.4 \times 10^{11}\ 1/\text{cm}^3$

(D) $7.0 \times 10^{17}\ 1/\text{cm}^3$

2. A diode can be formed by doping an otherwise pure crystal such that

(A) an n-type semiconductor is produced

(B) the crystal becomes an intrinsic semiconductor

(C) one half of the crystal is p-type and the other half is n-type

(D) a p-type semiconductor is produced

3. An ideal diode and two resistors are arranged in a series circuit as shown. The load resistor, R_L, dissipates 25 W.

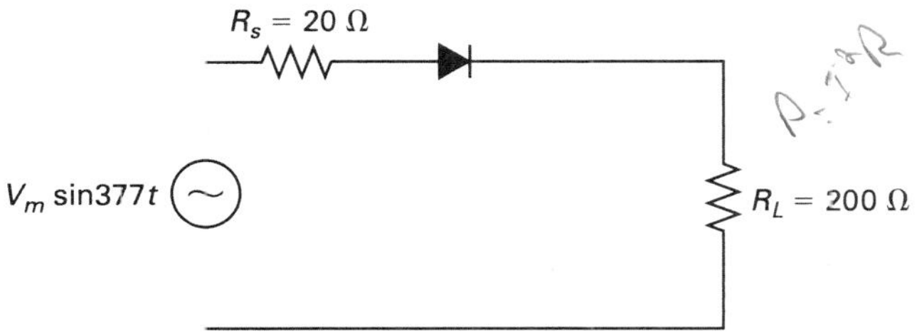

Most nearly, what is the magnitude, V_m, of the supply voltage?

(A) 100 V

(B) 110 V

(C) 140 V

(D) 160 V

4. A JFET is used in an amplifier circuit. The JFET operates in the saturation region. The conductivity factor is $6.5 \times 10^{-4}\ \text{A/V}^2$, and the transconductance is 2×10^{-3} S.

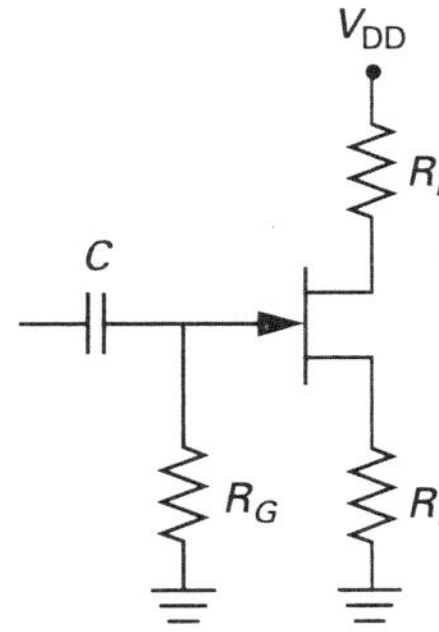

Most nearly, the drain current is

(A) 0.5 mA

(B) 1.0 mA

(C) 1.3 mA

(D) 1.5 mA

5. A black box is reverse-engineered and is found to contain the circuit shown.

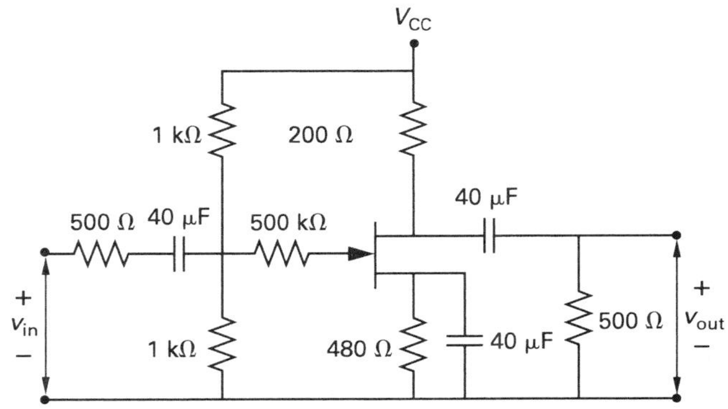

Most likely, what is the purpose of the capacitors in the circuit?

(A) to speed up the voltage propagation

(B) to isolate the DC biasing circuit from the AC input and output

(C) to suppress noise form the DC voltage source V_{CC}

(D) to increase the amplification

6. A BJT used in a circuit as shown has a common emitter current gain of 100 and a DC base-to-emitter voltage of 0.6 V. Most nearly, the output voltage of the circuit is

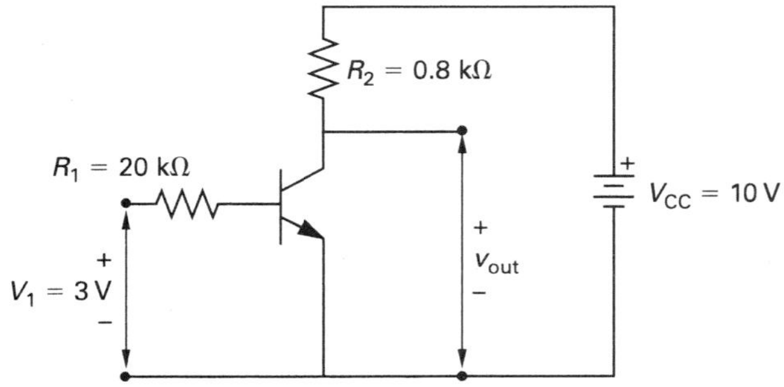

(A) 0.4 V

(B) 2 V

(C) 3 V

(D) 10 V

7. An *npn* BJT at 300K has a saturation current of 1 pA. Most nearly, the base-emitter voltage needed to create a collector current of 556 mA is

(A) 18 mV

(B) 0.70 V

(C) 22 V

(D) 27 V

8. An *npn* BJT has a common emitter current gain of 80 and a base current of 134 μA. Most nearly, the emitter current is

(A) 1.20 mA

(B) 5.40 mA

(C) 10.6 mA

(D) 10.9 mA

9. A *p*-channel JFET is biased by two ideal voltage sources, as shown. The pinch-off voltage for the JFET is –6.0 V, and the drain-to-source saturation current is 34.5 mA.

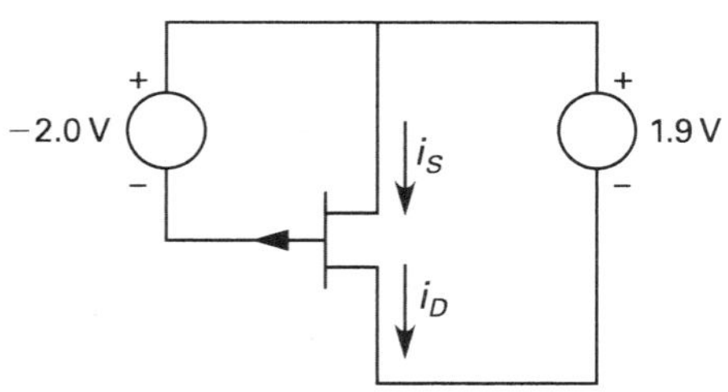

Most nearly, the drain current is

(A) 11 mA

(B) 15 mA

(C) 35 mA

(D) 170 mA

10. A *p*-channel JFET is operating in the saturation region. The drain-to-source current is 500 mA, the pinch-off voltage is 2.5 V, and the drain current is 34.5 mA. Most nearly, the small-signal transconductance is

(A) 0.24 mS

(B) 42 mS

(C) 110 mS

(D) 150 mS

11. An enhancement-mode NMOS MOSFET has a conductivity factor of 4 mA/V^2 and a threshold voltage of 0.7 V. The MOSFET is operating in the saturation region. Most nearly, the gate-to-source voltage needed to produce a drain current of 23.5 mA is

(A) 1.7 V

(B) 2.5 V

(C) 3.1 V

(D) 4.1 V

12. A silicon *pn* diode at room temperature has a thermal voltage of 0.025 V. Most nearly, the applied diode voltage needed to make the diode current equal 90% of the diode saturation current is

(A) 0.016 V

(B) 0.16 V

(C) 0.50 V

(D) 0.70 V

13. The JFET used in the circuit shown operates in the saturation region. The JFET's drain-to-source saturation current is 18 mA, and its pinch-off voltage is –3 V.

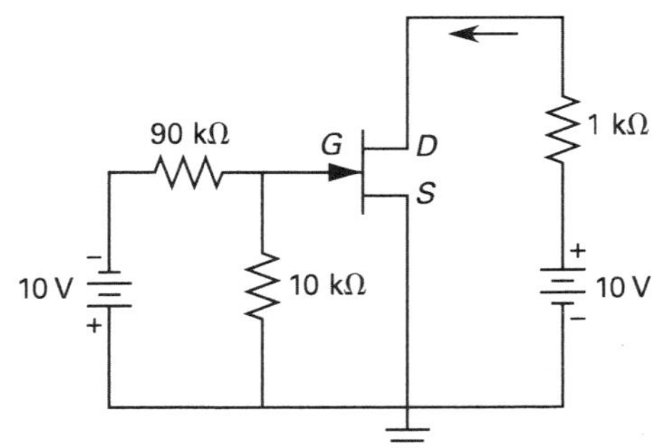

Most nearly, the power dissipated in the JFET is

(A) 0.016 W

(B) 0.16 W

(C) 1.6 W

(D) 16 W

14. The JFET shown is operating in the saturation region. The drain supply voltage is 5 V, the saturation resistance is 1 kΩ, the drain-to-source saturation current is 2 mA, and the pinch-off voltage is −2 V.

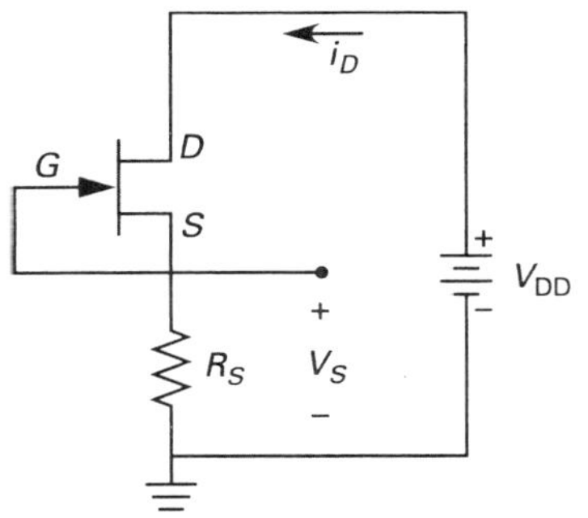

Most nearly, what is the gate-to-drain voltage?

(A) −3 V

(B) −2 V

(C) 2 V

(D) 0 V

15. The BJT incorporated into the circuit shown has a base-emitter voltage of 0.7 V and a common emitter current gain of 100.

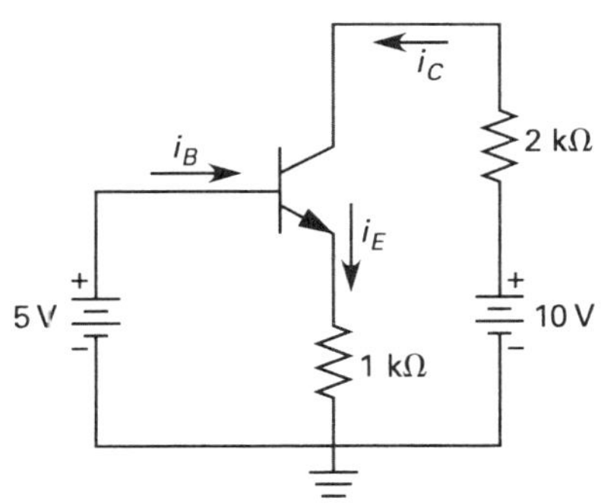

Most nearly, the base current is

(A) 0.043 mA

(B) 0.43 mA

(C) 0.75 mA

(D) 7.5 mA

16. The n-channel JFET in the circuit shown is operating in the saturation region. It has a pinch-off voltage of −4 V, a drain current of 7.2 mA, and a gate-to-source voltage of 10 V. The drain-to-source voltage is 8 V.

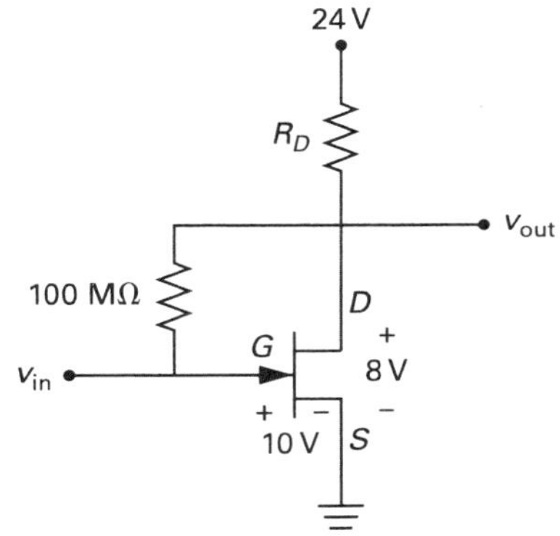

Most nearly, the drain resistance is

(A) 12.7 kΩ

(B) 13.0 kΩ

(C) 14.4 kΩ

(D) 15.0 kΩ

17. A JFET operating in the saturation region has a pinch-off voltage of −4 V, a drain current of 6.4 mA, and a gate-to-source voltage of 9 V. Most nearly, the transconductance is

(A) 9.9×10^{-4} S

(B) 6.7×10^{-4} S

(C) 2.8×10^{-3} S

(D) 1.3×10^{-3} S

18. The n-channel enhancement MOSFET shown is operating in the saturation region. It has a threshold voltage of 4 V, a drain current of 7.2 mA, and a gate-to-source voltage of 10 V. The drain-to-source voltage is 8 V.

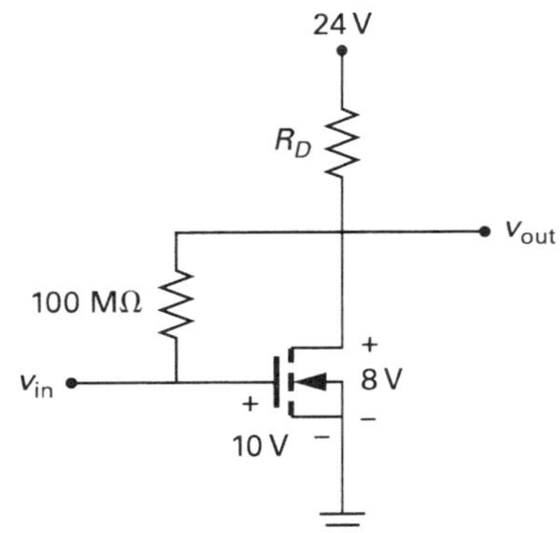

Electronics

Most nearly, what is the drain resistance?

(A) 2.7 kΩ

(B) 3.0 kΩ

(C) 4.4 kΩ

(D) 5.0 kΩ

SOLUTIONS

1. The absolute equilibrium temperature is

$$T = 50°\text{C} + 273.15° = 323.15\text{K}$$

Use the equation for the contact potential of a *pn* junction, and solve for the acceptor concentration.

$$V_0 = \frac{kT}{q}\ln\frac{N_aN_d}{n_i^2}$$

$$\begin{aligned} N_a &= \left(\frac{n_i^2}{N_d}\right)e^{V_0q/kT} \\ &= \left(\frac{\left(1.5\times 10^{10}\ \frac{1}{\text{cm}^3}\right)^2}{7.0\times 10^{17}\ \frac{1}{\text{cm}^3}}\right)\exp\frac{(0.6\ \text{V})(1.602\times 10^{-19}\ \text{C})}{(1.38\times 10^{-23}\ \text{J/K})(323.15\text{K})} \\ &= 7.4\times 10^{11}\ \text{cm}^{-3} \end{aligned}$$

The answer is (C).

2. A diode is a junction of *n*-type and *p*-type semiconductors. Option C is correct.

The answer is (C).

3. The circuit is a half-wave rectifier.

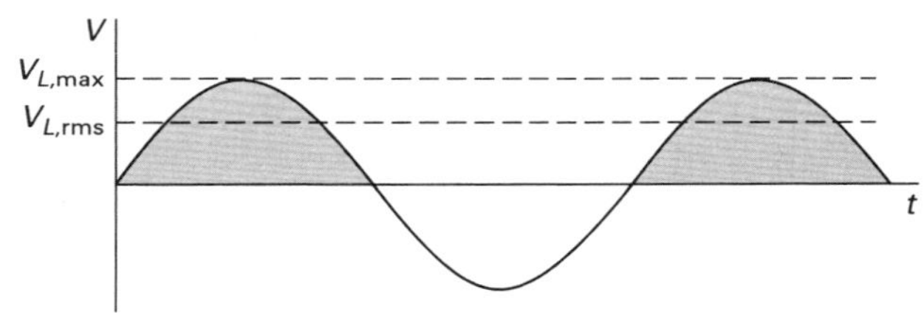

The power drawn by the load, P_L, is proportional to the rms value of the voltage. For a half-wave rectified sinusoid, the rms value is equal to one-half of the maximum value.

$$V_{L,\text{rms}} = \frac{V_{L,\text{max}}}{2}$$

Electronics

For a purely resistive load, power is

$$\begin{aligned}P_L &= I_{L,\text{rms}}V_{L,\text{rms}}\\ &= \left(\frac{V_{L,\text{rms}}}{200\ \Omega}\right)V_{L\ \text{rms}}\\ &= \frac{V_{L,\text{rms}}^2}{200\ \Omega}\\ &= \frac{\left(\frac{V_{L,\max}}{2}\right)^2}{200\ \Omega}\\ &= \frac{V_{L,\max}^2}{800\ \Omega}\end{aligned}$$

The load resistor dissipates 25 W, so

$$\begin{aligned}25\ \text{W} &= \frac{V_{L,\max}^2}{800\ \Omega}\\ V_{L,\max} &= \sqrt{(25\ \text{W})(800\ \Omega)}\\ &= 141.4\ \text{V}\quad [\text{at the load}]\end{aligned}$$

The shunt resistance creates a voltage divider.

$$\begin{aligned}V_{L,\max} &= V_m\left(\frac{R_L}{R_S+R_L}\right) = V_m\left(\frac{200\ \Omega}{20\ \Omega+200\ \Omega}\right)\\ &= 0.909\,V_m\\ V_m &= \frac{V_{L,\max}}{0.909} = \frac{141.4\ \text{V}}{0.909}\\ &= 155.6\ \text{V}\quad (160\ \text{V})\end{aligned}$$

The answer is (D).

4. Use the equation for JFET transconductance, and solve for the drain current, I_D.

$$\begin{aligned}g_m &= \frac{2\sqrt{I_{DSS}I_D}}{|V_p|}\\ I_D &= \frac{g_m^2V_p^2}{4I_{DSS}}\end{aligned}$$

From the equation for the drain-to-source saturation current, $I_{DSS} = KV_p^2$, so

$$\begin{aligned}I_D &= \frac{g_m^2V_p^2}{4KV_p^2} = \frac{g_m^2}{4K}\\ &= \frac{(2\times 10^{-3}\ \text{S})^2}{(4)\left(6.5\times 10^{-4}\ \frac{\text{A}}{\text{V}^2}\right)}\\ &= 1.54\times 10^{-3}\ \text{A}\quad (1.5\ \text{mA})\end{aligned}$$

The answer is (D).

5. The capacitors allow the DC bias on the transistor to operate the transistor in the desired range without the DC voltage affecting the input or the output. The purpose of the capacitors, then, is most likely to isolate the DC biasing circuit from the AC input and output.

The answer is (B).

6. Use Ohm's law to find the base current.

$$\begin{aligned}V &= IR\\ I &= \frac{V}{R}\\ i_B &= \frac{V_1 - V_{BE}}{R_1} = \left(\frac{3\ \text{V}-0.6\ \text{V}}{20\times 10^3\ \Omega}\right)\left(1000\ \frac{\text{mA}}{\text{A}}\right)\\ &= 0.12\ \text{mA}\end{aligned}$$

Find the collector current.

$$i_C = \beta i_B = (100)(0.12\,\text{mA}) = 12\,\text{mA}$$

Find the output voltage.

$$\begin{aligned}V_{\text{out}} &= V_{CC} - i_CR_2\\ &= 10\ \text{V} - \left(\frac{12\ \text{mA}}{1000\ \frac{\text{mA}}{\text{A}}}\right)\left((0.8\ \text{k}\Omega)\left(1000\ \frac{\Omega}{\text{k}\Omega}\right)\right)\\ &= 0.4\ \text{V}\end{aligned}$$

The answer is (A).

Electronics

7. The thermal voltage is

$$\begin{aligned} V_T &= \frac{kT}{q} \\ &= \frac{\left(1.38 \times 10^{-23}\ \frac{\text{J}}{\text{K}}\right)(300\text{K})}{1.6022 \times 10^{-19}\ \text{C}} \\ &= 0.0258\ \text{V} \end{aligned}$$

Use the equation for collector current. Solve for the base-emitter voltage, setting the collector current to 556 mA.

$$\begin{aligned} i_C &\approx I_S e^{(V_{BE}/V_T)} \\ V_{BE} &= V_T \ln\frac{i_C}{I_S} = (0.0258\ \text{V})\ln\frac{0.556\ \text{A}}{1 \times 10^{-12}\ \text{A}} \\ &= 0.70\ \text{V} \end{aligned}$$

The answer is (B).

8. Substitute the equation for collector current, i_C, into the equation for emitter current.

$$\begin{aligned} i_E &= i_B + i_C = i_B + \beta i_B = i_B(1+\beta) \\ &= (134 \times 10^{-6}\ \text{A})(1+80) \\ &= 10.9\ \text{mA} \end{aligned}$$

The answer is (D).

9. The drain current depends on the operating condition of the JFET. When a JFET is operating in the cutoff region, the gate-to-source voltage, v_{GS}, is less than the pinch-off voltage, V_p. In this case, however, the gate-to-source voltage, -2.0 V, is greater than the pinch-off voltage, -6.0 V, so the JFET is not operating in the cutoff region.

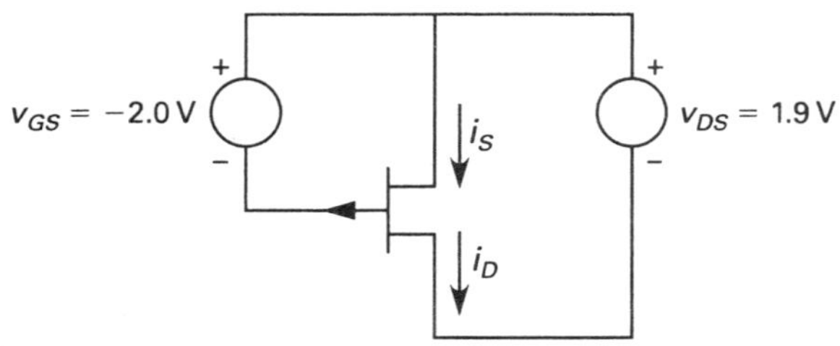

When a JFET is operating in the triode region, both the gate-to-source voltage and the gate-to-drain voltage, v_{GD}, are greater than the pinch-off voltage. Find the gate-to-drain voltage by subtracting the drain-to-source voltage from the gate-to-source voltage.

$$v_{GD} = v_{GS} - v_{DS} = -2.0\ \text{V} - 1.9\ \text{V} = -3.9\ \text{V}$$

Both the gate-to-drain voltage of -3.9 V and the gate-to-source voltage of -2.0 V are greater than the pinch-off voltage of -6.0 V, so the JFET is operating in the triode region. Use the JFET performance equation for the triode region to find the drain current.

$$\begin{aligned} i_D &= (I_{DSS}/V_p^2)\left[2v_{DS}(v_{GS} - V_p) - v_{DS}^2\right] \\ &= \left(\frac{34.5\ \text{mA}}{(-6.0\ \text{V})^2}\right)\left((2)(1.9\ \text{V})(-2.0 - (-6.0\ \text{V})) - (1.9\ \text{V})^2\right) \\ &= 11.0\ \text{mA} \quad (11\ \text{mA}) \end{aligned}$$

The answer is (A).

10. The small-signal transconductance of a JFET operating in the saturation region is

$$\begin{aligned} g_m &= \frac{2\sqrt{I_{DSS}I_D}}{|V_p|} = \frac{2\sqrt{(500 \times 10^{-3}\ \text{A})(34.5 \times 10^{-3}\ \text{A})}}{|2.5\ \text{V}|} \\ &= 0.105\ \text{S} \quad (110\ \text{mS}) \end{aligned}$$

The answer is (C).

11. Use the drain current equation for a MOSFET operating in the saturation region, and solve for the gate-to-source voltage.

$$\begin{aligned} i_D &= K(v_{GS} - V_t)^2 \\ v_{GS} &= \sqrt{\frac{i_D}{K}} + V_t = \sqrt{\frac{23.5\ \text{mA}}{4\ \frac{\text{mA}}{\text{V}^2}}} + 0.7\ \text{V} \\ &= 3.12\ \text{V} \quad (3.1\ \text{V}) \end{aligned}$$

The answer is (C).

12. For silicon, the emission coefficient, η, is 1. Use Shockley's equation, and set the diode current, i_D, equal to 90% of the saturation current, I_s.

$$\begin{aligned} i_D &\approx I_s\left[e^{(v_D/\eta V_T)} - 1\right] \\ 0.9I_s &= I_s(e^{v_D/0.025\ \text{V}} - 1) \\ 1.9 &= e^{v_D/0.025\ \text{V}} \\ \ln 1.9 &= \frac{v_D}{0.025\ \text{V}} \\ v_D &= (\ln 1.9)(0.025\ \text{V}) \\ &= 0.01605\ \text{V} \quad (0.016\ \text{V}) \end{aligned}$$

The answer is (A).

13. The JFET is as shown.

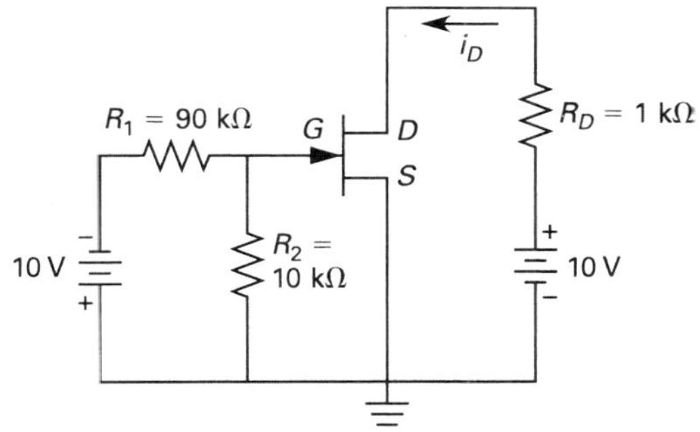

Use the voltage divider concept on the left side of the circuit. The battery polarity is negative, so use a negative value for the voltage.

$$\begin{aligned} v_{GS} &= \left(\frac{R_2}{R_1 + R_2}\right)V_1 \\ &= \left(\frac{10\ \text{k}\Omega}{90\ \text{k}\Omega + 10\ \text{k}\Omega}\right)(-10\ \text{V}) \\ &= -1.0\ \text{V} \end{aligned}$$

The drain current is

$$\begin{aligned} i_D &= I_{DSS}(1 - v_{GS}/V_p)^2 \\ &= (18\ \text{mA})\left(1 - \frac{-1.0\ \text{V}}{-3\ \text{V}}\right)^2 \\ &= 8\ \text{mA} \end{aligned}$$

The source-to-drain voltage is

$$\begin{aligned} v_{DS} &= R_2 - i_D R_D \\ &= 10\ \text{V} - \left(\frac{8\ \text{mA}}{1000\ \frac{\text{mA}}{\text{A}}}\right)\left((1\ \text{k}\Omega)\left(1000\ \frac{\Omega}{\text{k}\Omega}\right)\right) \\ &= 2\ \text{V} \end{aligned}$$

The power dissipated in the JFET is

$$\begin{aligned} P = v_{DS} i_D &= (2\ \text{V})\left(\frac{8\ \text{mA}}{1000\ \frac{\text{mA}}{\text{A}}}\right) \\ &= 0.016\ \text{W} \end{aligned}$$

The answer is (A).

14. Because the gate and the source are tied together, the gate-to-source voltage, v_{GS}, is zero. Find the drain current using the equation for JFET performance in the saturation region.

$$\begin{aligned} i_D &= I_{DSS}(1 - v_{GS}/V_p)^2 \\ &= (2\ \text{mA})\left(1 - \frac{0\ \text{V}}{V_p}\right)^2 \\ &= 2\ \text{mA} \end{aligned}$$

Use Ohm's law to find the saturation voltage.

$$\begin{aligned} V_S &= i_D R_S \\ &= \left(\frac{2\ \text{mA}}{1000\ \frac{\text{mA}}{\text{A}}}\right)\left((1\ \text{k}\Omega)\left(1000\ \frac{\Omega}{\text{k}\Omega}\right)\right) \\ &= 2\ \text{V} \end{aligned}$$

The drain-to-source voltage is

$$v_{DS} = V_{DD} - V_S = 5\ \text{V} - 2\ \text{V} = 3\ \text{V}$$

The gate-to-drain voltage is

$$v_{GD} = -v_{DS} = -3\ \text{V}$$

The answer is (A).

15. The circuit is as shown.

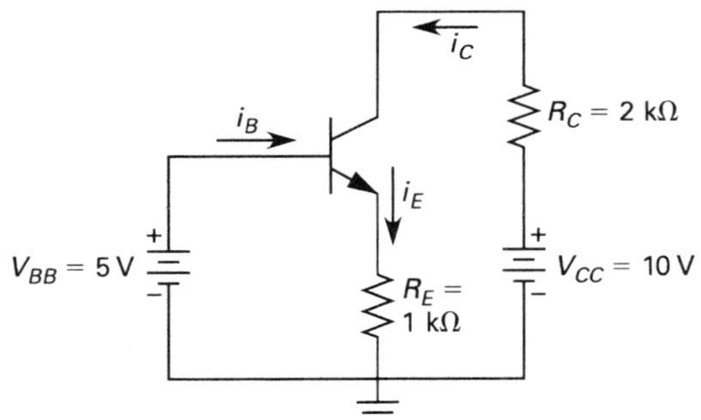

Write Kirchhoff's voltage law for the base-emitter loop, and solve for the emitter current.

$$V_{BB} = V_{BE} + i_E R_E$$

$$\begin{aligned} i_E &= \frac{V_{BB} - V_{BE}}{R_E} = \frac{(5\ \text{V} - 0.7\ \text{V})\left(1000\ \frac{\text{mA}}{\text{A}}\right)}{(1\ \text{k}\Omega)\left(1000\ \frac{\Omega}{\text{k}\Omega}\right)} \\ &= 4.3\ \text{mA} \end{aligned}$$

Electronics

Substitute the equation for collector current, I_C, into the equation for emitter current, I_E.

$$i_E = i_B + i_C = i_B + \beta i_B = (1+\beta)i_B$$

$$i_B = \frac{i_E}{1+\beta} = \frac{4.3\text{ mA}}{1+100} = 0.0426\text{ mA} \quad (0.043\text{ mA})$$

The answer is (A).

16. The circuit is as shown.

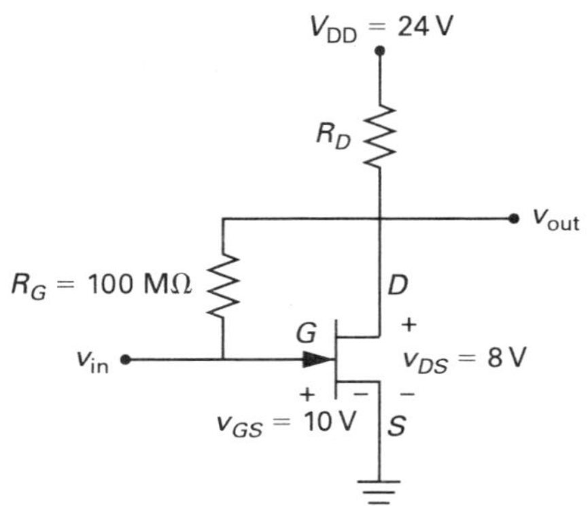

Use the equation for JFET operation in the saturation region, and solve for the drain-to-source saturation current.

$$i_D = I_{DSS}(1 - v_{GS}/V_p)^2$$

$$I_{DSS} = \frac{i_D}{\left(1 - \frac{v_{GS}}{V_p}\right)^2} = \frac{0.0072\text{ A}}{\left(1 - \frac{10\text{ V}}{-4\text{ V}}\right)^2} = 1.37 \times 10^{-4}\text{ A}$$

Because the current into the gate is minimal, the drop across R_G is insignificant. The gate-to-source voltage and the drain-to-source voltage will be approximately equal.

$$v_{GS} \approx v_{DS} = 8\text{ V}$$

The drain-to-source current is

$$i_D = I_{DSS}(1 - v_{GS}/V_p)^2 = (1.37 \times 10^{-4}\text{ A})\left(1 - \frac{8\text{ V}}{-4\text{ V}}\right)^2 = 1.23 \times 10^{-3}\text{ A}$$

Write Kirchhoff's voltage law around the output circuit, and solve for R_D.

$$V_{DD} - i_D R_D - v_{DS} = 0$$

$$R_D = \frac{V_{DD} - v_{DS}}{i_D} = \frac{24\text{ V} - 8\text{ V}}{1.23 \times 10^{-3}\text{ A}} = 13.0 \times 10^3\ \Omega \quad (13.0\text{ k}\Omega)$$

The answer is (B).

17. Use the equation for JFET operation in the saturation region, and solve for the drain-to-source saturation current.

$$i_D = I_{DSS}(1 - v_{GS}/V_p)^2$$

$$I_{DSS} = \frac{i_D}{\left(1 - \frac{v_{GS}}{V_p}\right)^2} = \frac{6.4 \times 10^{-3}\text{ A}}{\left(1 - \frac{9\text{ V}}{-4\text{ V}}\right)^2} = 6.06 \times 10^{-4}\text{ A}$$

The transconductance is

$$g_m = \frac{2\sqrt{I_{DSS} I_D}}{|V_p|} = \frac{2\sqrt{(6.06 \times 10^{-4}\text{ A})(6.4 \times 10^{-3}\text{ A})}}{|-4\text{ V}|} = 9.85 \times 10^{-4}\text{ S} \quad (9.9 \times 10^{-4}\text{ S})$$

The answer is (A).

18. Use the equation for MOSFET operation in the saturation region, and solve for the conductivity factor.

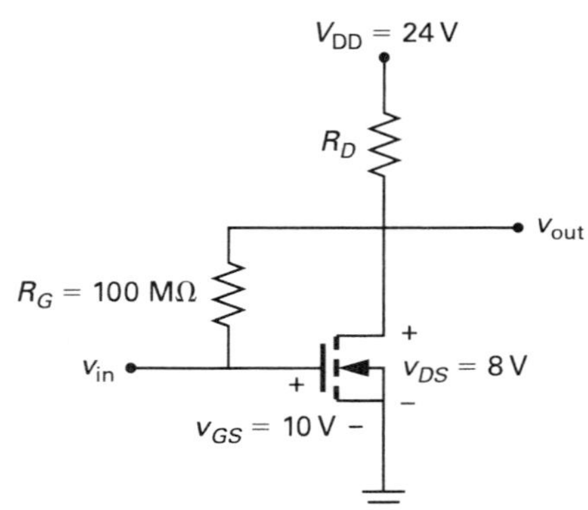

$$i_D = K(v_{GS} - V_t)^2$$

$$K = \frac{i_D}{(v_{GS} - V_t)^2} = \frac{0.0072\text{ A}}{(10\text{ V} - 4\text{ V})^2} = 0.0002\text{ A/V}^2$$

Electronics

Because the current into the gate is minimal, the drop across R_G is insignificant. The gate-to-source voltage and the drain-to-source voltage will be approximately equal.

$$v_{GS} \approx v_{DS} = 8\text{ V}$$

Because the current into the base is insignificant, the current through R_D will be equal to the drain-to-source current. In the saturation region, the drain-source current is

$$\begin{aligned} i_D &= K(v_{GS} - V_t)^2 \\ &= \left(0.0002\ \frac{\text{A}}{\text{V}^2}\right)(8\text{ V} - 4\text{ V})^2 \\ &= 3.2 \times 10^{-3}\text{ A} \end{aligned}$$

Write Kirchhoff's voltage law for the output circuit and solve for R_D.

$$\begin{aligned} V_{\text{DD}} - i_D R_D - V_{DS} &= 0 \\ R_D &= \frac{V_{\text{DD}} - V_{DS}}{i_D} = \frac{24\text{ V} - 8\text{ V}}{3.2 \times 10^{-3}\text{ A}} \\ &= 5.0 \times 10^3\ \Omega \quad (5.0\text{ k}\Omega) \end{aligned}$$

The answer is (D).

Electronics

31 Amplifiers

PRACTICE PROBLEMS

1. The ideal op amps are cascaded as shown.

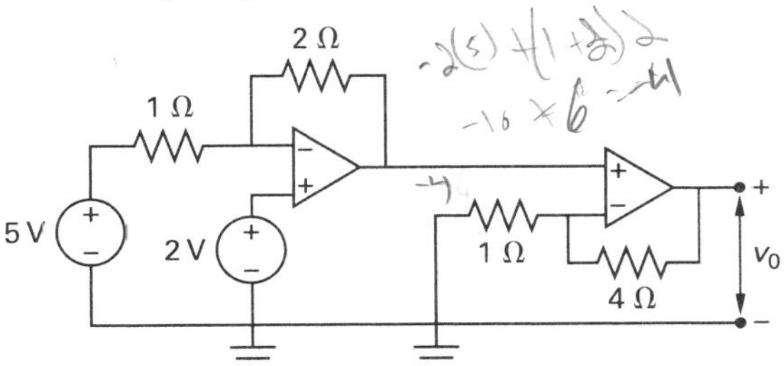

What is most nearly the output voltage?

(A) −20 V

(B) 2.5 V

(C) 4.0 V

(D) 16 V

2. An amplifier with a gain of 2.0 is shown. The input signal is $v_{in}(t) = (5 \text{ V})\sin((60\pi)t + 28°)$. The source voltage is 8.0 V.

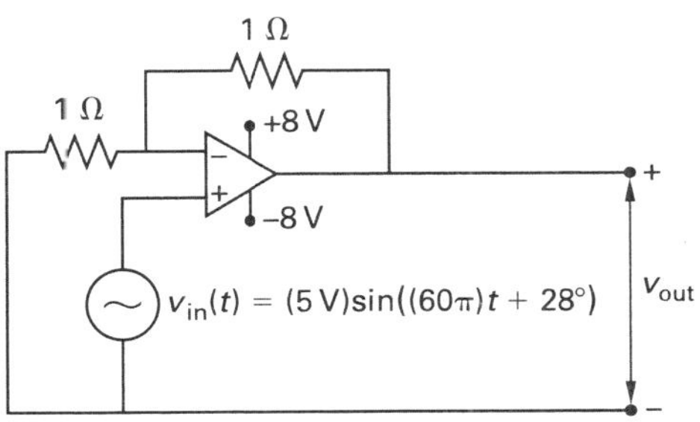

What is most nearly the output voltage at $t = 0.025$ s?

(A) −8.8 V

(B) −8.0 V

(C) −4.4 V

(D) 8.0 V

3. An op amp circuit is shown. The non-inverting input is connected to a 12 V DC source. The op amp voltage gain, A, is 100 000, the input resistance, R_{in}, is 1.0 MΩ, and the output resistance, R_{out}, is 20 Ω.

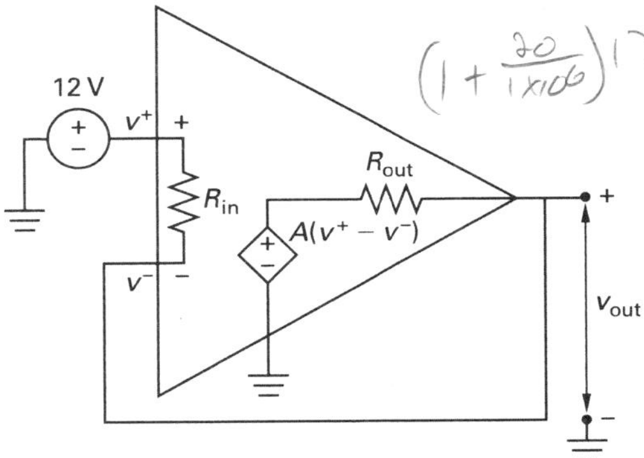

What is most nearly the gain of the circuit?

(A) 1.0

(B) 5.0

(C) 10

(D) 20

4. A half-rectified sine wave is the input to an amplifier with an amplification factor of −2.

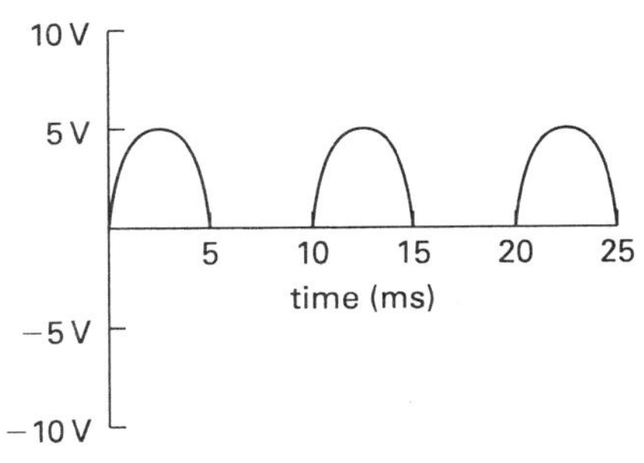

Electronics

Which graph most nearly shows the waveform of the output?

(A)

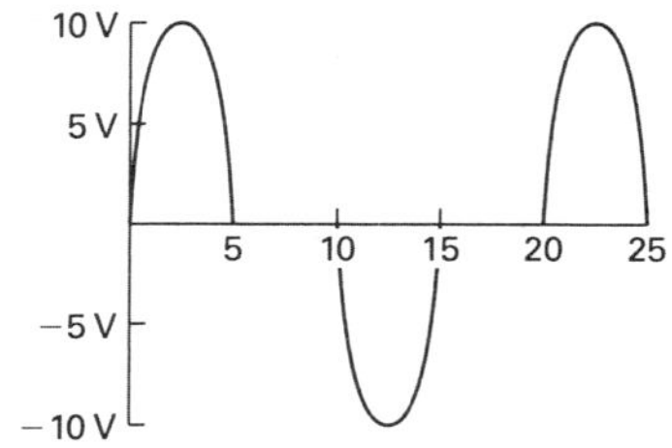

(B)

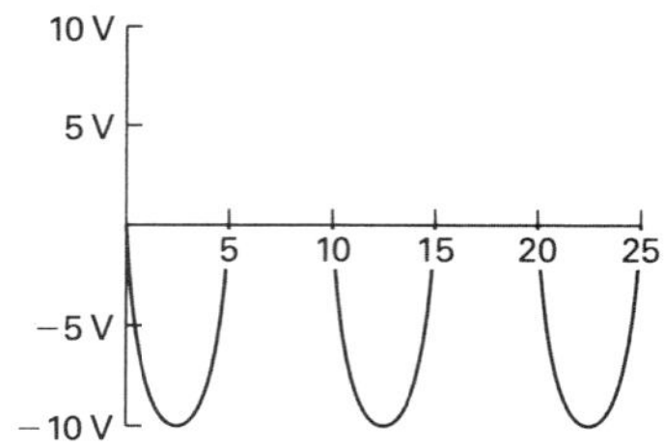

(C)

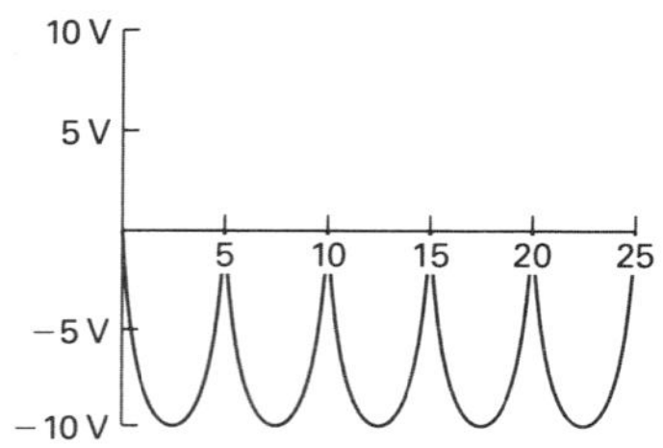

(D)

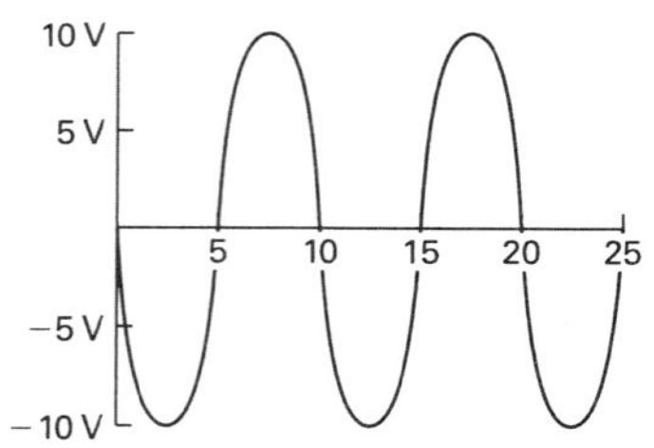

5. During testing of an operational amplifier, signals of equal magnitude are applied to the two terminals (i.e., $v_1 = -v_2$). What gain is measured?

(A) A_{cm}

(B) A_{id}

(C) A_V

(D) G_P

6. An ideal operational amplifier is shown. The voltage at the inverting terminal, v_2, is 0.002 V, and the voltage at the noninverting terminal, v_1, is 0.001 V. The output voltage is −10 V. Most nearly, what is the gain of the operational amplifier?

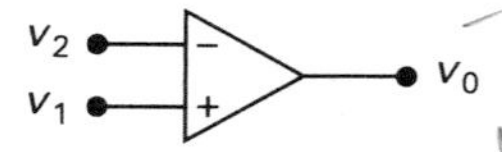

(A) 1000

(B) 10,000

(C) 100,000

(D) 1,000,000

7. An ideal operational amplifier is used in a voltage amplification circuit as shown.

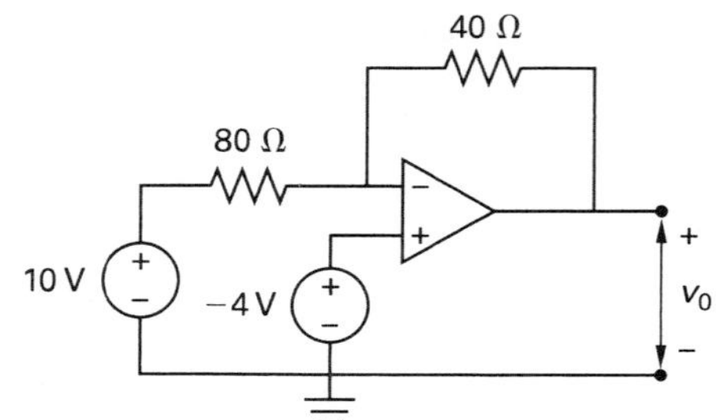

Most nearly, what is the output voltage?

(A) −14 V

(B) −11 V

(C) −1 V

(D) 6 V

8. An ideal operational amplifier is used in a voltage amplification circuit as shown.

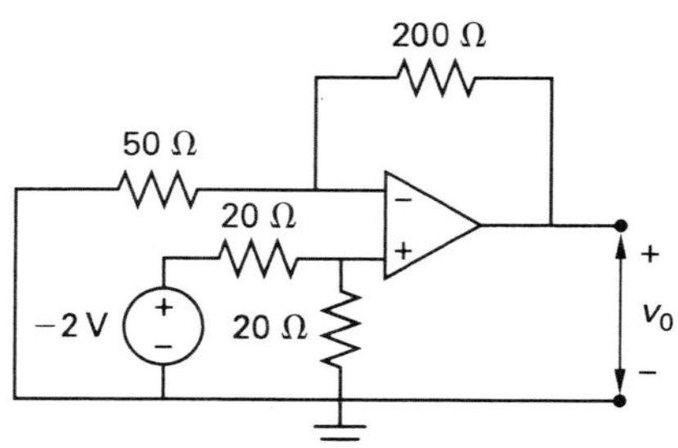

Most nearly, what is the output voltage?

(A) −10 V

(B) −5 V

(C) 5 V

(D) 10 V

9. The voltage gain in the ideal operational amplifier circuit shown is −200.

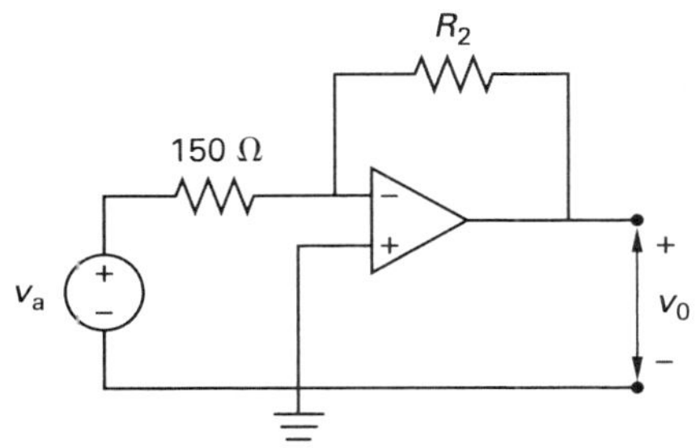

Most nearly, what is the value of R_2?

(A) 3 kΩ

(B) 30 kΩ

(C) 100 kΩ

(D) 300 kΩ

10. The inverting input of a nonideal operational amplifier is at 0.001 V. The noninverting voltage is 0.002 V, and the output voltage is 10.01 V. The differential gain of the operational amplifier is 10,000. Most nearly, what is the common-mode rejection ratio?

(A) 64 dB

(B) 74 dB

(C) 98 dB

(D) 134 dB

11. For a differential amplifier, the ratio of the first emitter current to the second emitter current is 2.0. The voltage difference between the first base current and the second base current is 0.02 V. Most nearly, what is the temperature of operation?

(A) 335K

(B) 345K

(C) 465K

(D) 485K

SOLUTIONS

1. The circuit can be split into two parts, where the output voltage of the configuration in the left box, $v_{0,1}$, is the input to the configuration in the right box, and the output voltage of the configuration in the right box, $v_{0,2}$, is equal to the total output voltage of the circuit. The configuration in the left box acts as an inverting amplifier on v_a and as a noninverting amplifier on v_b.

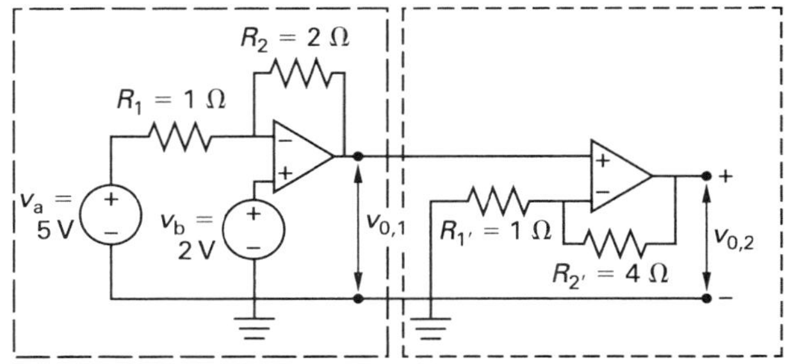

The output voltage of the configuration in the left box, $v_{0,1}$, is

$$\begin{aligned} v_{0,1} &= -\frac{R_2}{R_1}v_a + \left(1 + \frac{R_2}{R_1}\right)v_b \\ &= -\left(\frac{2\ \Omega}{1\ \Omega}\right)(5\ \text{V}) + \left(1 + \frac{2\ \Omega}{1\ \Omega}\right)(2\ \text{V}) \\ &= -4\ \text{V} \end{aligned}$$

Since $v_{0,1}$ is not 0 and is the sole input to the configuration in the right box, the configuration in the right box is a noninverting amplifier. The output voltage for the configuration in the right box, $v_{0,2}$, is

$$\begin{aligned} v_{0,2} &= \left(1 + \frac{R_{2'}}{R_{1'}}\right)v_{0,1} = \left(1 + \frac{4\ \Omega}{1\ \Omega}\right)(-4\ \text{V}) \\ &= -20\ \text{V} \end{aligned}$$

The answer is (A).

2. The output signal can be drawn as shown.

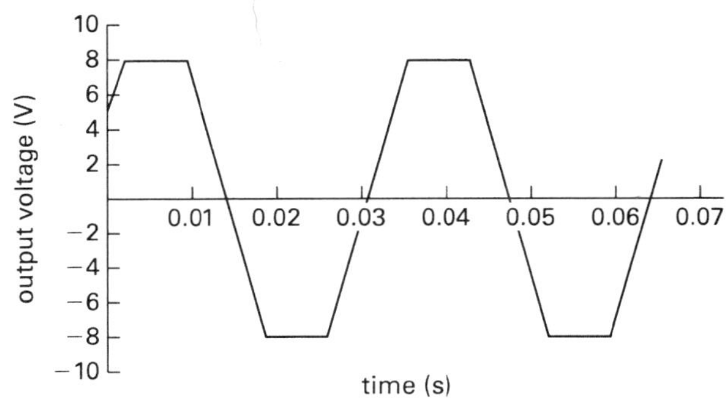

Electronics

Since the gain is 2.0 and $v_{out} = Av_{in}(t)$, the output voltage at 0.025 s is

$$\begin{aligned} v_{out} &= Av_{in}(t) = (2.0)\big((5\ \text{V})\sin((60\pi)t + 28°)\big) \\ &= (2.0)\left((5\ \text{V})\sin\left(60\pi\left(\frac{360°}{2\pi\ \text{rad}}\right)(0.025\ \text{s}) + 28°\right)\right) \\ &= -8.83\ \text{V} \end{aligned}$$

The output voltage cannot be less than -8.0 V, so the output voltage is -8.0 V.

The answer is (B).

3. Use loop analysis to find the equations for the 12 V source and v_{out}. Use the relationships shown for the path of the current.

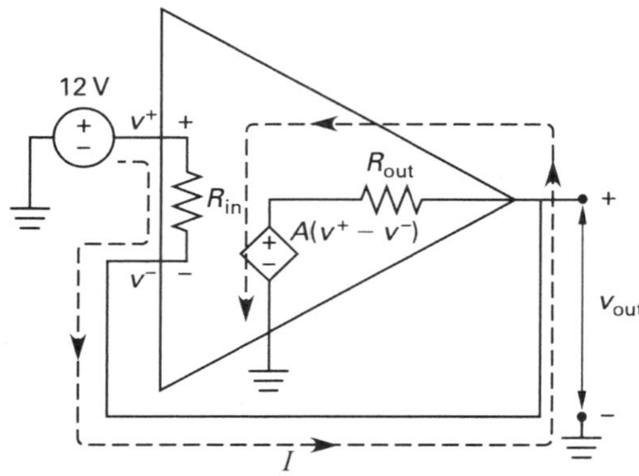

$$\begin{aligned} 12\ \text{V} &= IR_{in} + IR_{out} + A(v^+ - v^-) \\ v_{out} &= A(v^+ - v^-) + IR_{out} \end{aligned}$$

Since the voltage across v^+ and v^- is IR_{in}, the equations become

$$\begin{aligned} 12\ \text{V} &= IR_{in} + IR_{out} + AIR_{in} \\ v_{out} &= AIR_{in} + IR_{out} \end{aligned}$$

Divide the equation for v_{out} by the equation for the 12 V source to find the gain of the configuration.

$$\begin{aligned} A_{configuration} &= \frac{v_{out}}{12\ \text{V}} = \frac{AIR_{in} + IR_{out}}{IR_{in} + IR_{out} + AIR_{in}} \\ &= \frac{1}{1 + \dfrac{R_{in}}{R_{out} + AR_{in}}} \\ &= \frac{1}{1 + \dfrac{1 \times 10^6\ \Omega}{20\ \Omega + (1 \times 10^5)(1 \times 10^6\ \Omega)}} \\ &= 1.0 \end{aligned}$$

The answer is (A).

4. The voltage input to the amplifier between 0 ms and 5 ms, between 10 ms and 15 ms, and between 20 ms and 25 ms is the same, so the output voltage at those three intervals will be the same. The maximum input voltage at those intervals is 5 V, so the minimum output voltage at those intervals is

$$\begin{aligned} A &= \frac{v_{out}}{v_{in}} \\ v_{out} &= Av_{in} = (-2)(5\ \text{V}) \\ &= -10\ \text{V} \end{aligned}$$

Since the amplification factor is negative and all nonzero inputs in the given waveform are positive, all nonzero outputs must be negative. Option A shows positive output voltages between 0 ms and 5 ms and between 20 ms and 25 ms, and option D shows positive output voltages between 5 ms and 10 ms and between 15 ms and 20 ms, so option A and option D do not show the waveform of the output. The input voltage is 0 V between 5 ms and 10 ms and between 15 ms and 20 ms, resulting in a 0 V output for these intervals. Option C shows outputs during these intervals, so option C does not show the waveform of the output. Option B shows negative outputs at the intervals where the input is positive and 0 V outputs at the intervals where the input is 0 V, and the output voltage is never less than -10 V. Option B shows the waveform of the output.

The answer is (B).

5. The differential-mode voltage is

$$v_{id} = v_2 - v_1$$

Therefore, with $v_1 = -v_2$,

$$v_{id} = v_1 + v_1 = 2v_1$$

The common-mode voltage is

$$v_{icm} = \frac{v_1 + v_2}{2} = \frac{v_1 + (-v_1)}{2} = 0\ \text{V}$$

Therefore, only the difference is being amplified, and the gain is A_{id}.

The answer is (B).

6. Use the equation for output voltage, and rearrange to solve for the gain.

$$\begin{aligned} v_0 &= A(v_1 - v_2) \\ A &= \frac{v_0}{v_1 - v_2} = \frac{-10\ \text{V}}{0.001\ \text{V} - 0.002\ \text{V}} \\ &= 10{,}000 \end{aligned}$$

The answer is (B).

Electronics

7. Use the equation for a two-source operational amplifier. The output voltage is

$$\begin{aligned} v_0 &= -\frac{R_2}{R_1}v_a + \left(1 + \frac{R_2}{R_1}\right)v_b \\ &= \left(-\frac{40\ \Omega}{80\ \Omega}\right)(10\ \text{V}) + \left(1 + \frac{40\ \Omega}{80\ \Omega}\right)(-4\ \text{V}) \\ &= -11\ \text{V} \end{aligned}$$

The answer is (B).

8. The voltage at the noninverting terminal is a voltage divider of the input voltage.

$$\begin{aligned} V_b = \left(\frac{R_1}{R_1 + R_2}\right)V &= \left(\frac{20\ \Omega}{20\ \Omega + 20\ \Omega}\right)(-2\ \text{V}) \\ &= -1\ \text{V} \end{aligned}$$

The inverting voltage is zero, so use the equation for a noninverting operational amplifier. The output voltage is

$$\begin{aligned} v_0 &= \left(1 + \frac{R_2}{R_1}\right)V_b \\ &= \left(1 + \frac{200\ \Omega}{50\ \Omega}\right)(-1\ \text{V}) \\ &= -5\ \text{V} \end{aligned}$$

The answer is (B).

9. The voltage gain is $v_0/v_a = -200$. The noninverting voltage is zero, so use the equation for the output voltage of an inverting operational amplifier, and rearrange to solve for R_2.

$$v_0 = -\frac{R_2}{R_1}v_a$$

$$\begin{aligned} R_2 = -\frac{v_0 R_1}{v_a} = -\left(\frac{v_0}{v_a}\right)R_1 &= -(-200)\left(\frac{150\ \Omega}{1000\ \frac{\Omega}{\text{k}\Omega}}\right) \\ &= 30\ \text{k}\Omega \end{aligned}$$

The answer is (B).

10. The differential input voltage is

$$v_{id} = v_1 - v_2 = 0.002\ \text{V} - 0.001\ \text{V} = 0.001\ \text{V}$$

The common-mode input voltage is

$$v_{icm} = (v_1 + v_2)/2 = \frac{0.002\ \text{V} + 0.001\ \text{V}}{2} = 0.0015\ \text{V}$$

Use the equation for output voltage, and rearrange to solve for the common-mode gain.

$$\begin{aligned} v_O &= Av_{id} + A_{cm}v_{icm} \\ A_{cm} &= \frac{v_O - Av_{id}}{v_{icm}} \\ &= \frac{10.01\ \text{V} - (10{,}000)(0.001\ \text{V})}{0.0015\ \text{V}} \\ &= 6.67 \end{aligned}$$

The common-mode rejection ratio in decibels is

$$\begin{aligned} \text{CMRR} &= 20\log_{10}\left(\frac{|A|}{|A_{cm}|}\right) = 20\log_{10}\frac{|10{,}000|}{|6.67|} \\ &= 63.5\ \text{dB} \quad (64\ \text{dB}) \end{aligned}$$

The answer is (A).

11. The ratio of emitter currents is

$$\begin{aligned} \frac{i_{E1}}{i_{E2}} &= e^{(v_{B1} - v_{B2})/V_T} \\ 2.0 &= e^{0.02\ \text{V}/V_T} \end{aligned}$$

Take the natural logarithm of both sides and solve for the thermal voltage.

$$\begin{aligned} \ln 2.0 &= \frac{0.02\ \text{V}}{V_T} \\ V_T &= \frac{0.02\ \text{V}}{\ln 2.0} \end{aligned}$$

Use the equation for thermal voltage and solve for the temperature.

$$\begin{aligned} V_T &= \frac{kT}{q} \\ T = \frac{V_T q}{k} &= \frac{\left(\frac{0.02\ \text{V}}{\ln 2.0}\right)(1.6022 \times 10^{-19}\ \text{C})}{1.38 \times 10^{-23}\ \frac{\text{J}}{\text{K}}} \\ &= 334.5\text{K} \quad (335\text{K}) \end{aligned}$$

The answer is (A).

Electronics

32 Measurement and Instrumentation

PRACTICE PROBLEMS

1. An analog signal is sampled and processed digitally. The filtered signal is sent through an output digital-to-analog (D/A) converter where the analog signal is reconstructed.

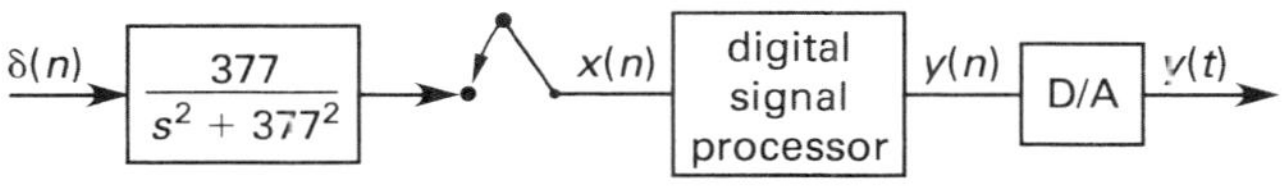

Most nearly, what is the minimum sampling frequency needed to accurately reconstruct the signal without aliasing?

(A) 60 Hz

(B) 120 Hz

(C) 190 Hz

(D) 380 Hz

2. Most nearly, what minimum sampling frequency will avoid aliasing with the analog signal, $x(t)$?

$$x(t) = \cos 100\pi t + \sin 200\pi t + \sin 60\pi t$$

(A) 30 Hz

(B) 60 Hz

(C) 200 Hz

(D) 400 Hz

3. It is desired to choose a resistance temperature detector (RTD) transducer for a measurement. The actual value of the temperature is not important because the initial temperature is known by other means, but the change in temperature as the test item is heated is important. Which statement is true about the selection of the RTD?

(A) The precision is important, and the accuracy is not important.

(B) The accuracy is important, and the precision is not important.

(C) Neither the accuracy nor the precision is important.

(D) Both accuracy and precision are important.

4. Using a gage factor of 2.1, what is most nearly the output sensitivity for the strain gauge bridge circuit shown?

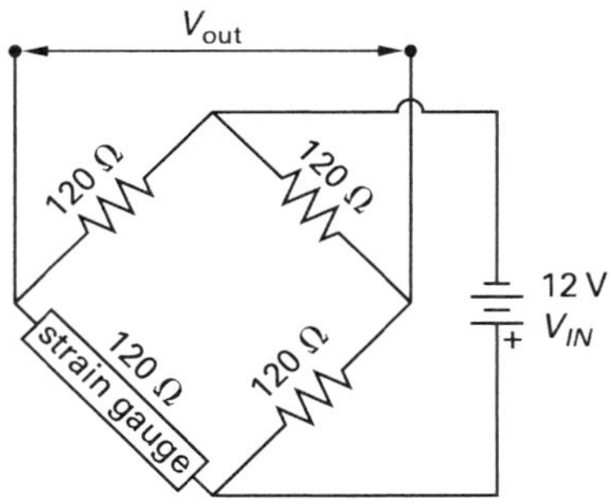

(A) 0.5 V·cm/cm

(B) 3.0 V·cm/cm

(C) 6.0 V·cm/cm

(D) 6.3 V·cm/cm

5. The output sensitivity of a strain gauge bridge circuit is 10 V·cm/cm. The output voltage is 10 mV. Most nearly, the strain is

(A) 0.0001 cm/cm

(B) 0.0010 cm/cm

(C) 0.0100 cm/cm

(D) 0.1000 cm/cm

6. Most nearly, what is the minimum sampling rate of a 4 kHz wideband signal to ensure accurate reconstruction of the signal?

(A) 4 kHz

(B) 7 kHz

(C) 8 kHz

(D) 13 kHz

SOLUTIONS

1. The signal, $x(t)$, generated by the oscillator can be determined by inspecting the Laplace transform table. The signal is

$$x(t) = \sin 377t = \sin 2\pi ft = \sin 2\pi 60t$$

To reconstruct $x(t)$ from $x(n)$ without aliasing, the sampling rate must be at least the Nyquist rate.

$$f_s \geq 2f_N = (2)(60\ \text{Hz}) = 120\ \text{Hz}$$

The answer is (B).

2. To avoid aliasing, the analog signal must be sampled at least at the Nyquist rate, which is twice the highest frequency contained in the signal. The highest frequency contained in the given signal is 200π rad/s or 100 Hz. Therefore, the minimum frequency that can be used to avoid aliasing is 200 Hz.

The answer is (C).

3. Precision is a measure of the repeatability of results, which is important for this measurement. The accuracy can have a significant bias, and the RTD would be acceptable for this measurement, so the accuracy is not important.

The answer is (A).

4. Find the change in resistance, ΔR.

$$GF = \frac{\Delta R/R}{\Delta L/L} = \frac{\Delta R/R}{\varepsilon}$$

$$\Delta R = \varepsilon(R)(GF)$$

Substitute the equation for ΔR into the equation for reference voltage.

$$\begin{aligned} V_0 &\approx \frac{\Delta R}{4R} \cdot V_{\text{IN}} \\ &= \left(\frac{\varepsilon(R)(GF)}{4R}\right) V_{\text{IN}} \\ \frac{V_0}{\varepsilon} &= \frac{V_{IN}(GF)}{4} = \frac{(12\ \text{V})(2.1)}{4} \\ &= 6.3\ \text{V·cm/cm} \end{aligned}$$

(Any linear unit can be used for strain.)

The answer is (D).

5. Use the definition of sensitivity to calculate the strain.

$$\text{sensitivity} = \frac{V_0}{\varepsilon}$$

$$\begin{aligned} \varepsilon &= \frac{V_0}{\text{sensitivity}} \\ &= \frac{10\ \text{mV}}{\left(10\ \frac{\text{V·cm}}{\text{cm}}\right)\left(1000\ \frac{\text{mV}}{\text{V}}\right)} \\ &= 0.0010\ \text{cm/cm} \end{aligned}$$

The answer is (B).

6. The minimum sampling rate is the Nyquist rate, which is given by

$$f_s \geq 2W = (2)(4\ \text{kHz}) = 8\ \text{kHz}$$

The answer is (C).

Electronics

33 Control Systems

PRACTICE PROBLEMS

1. The frequency response of a system to a sinusoidal input is given by

$$\frac{M}{\alpha} = \frac{X_2(j\omega)}{X_1(j\omega)}$$

By differentiation, the peak value of M, M_p, and the frequency at which it occurs, ω_r, are expressed in terms of the damping ratio, ζ, and natural frequency, ω_n.

$$M_p = \frac{1}{2\zeta\sqrt{1-\zeta^2}}$$

$$\omega_r = \omega_n\sqrt{1-2\zeta^2}$$

Which curve in the illustration shown best represents the frequency response?

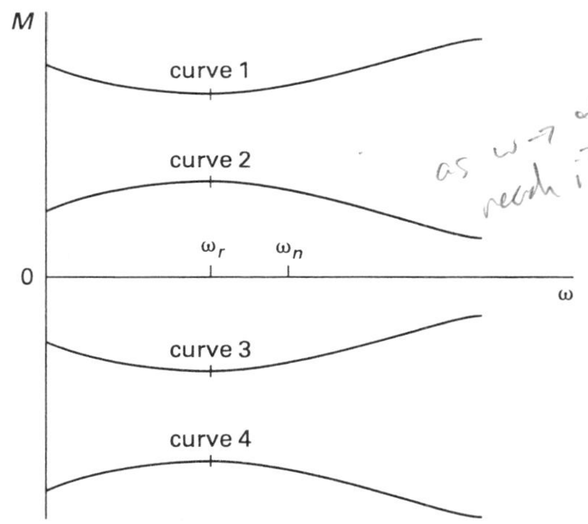

(A) curve 1

(B) curve 2

(C) curve 3

(D) curve 4

2. For the given block diagram, K is set to 16.

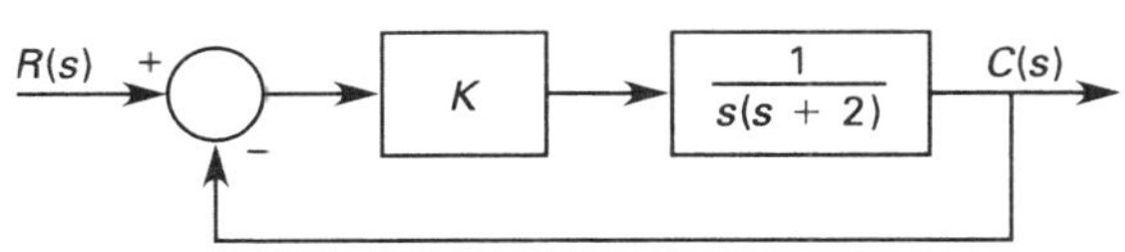

The damped natural frequency of oscillation in response to a step input, ω_d, is most nearly

(A) 3.9 rad/s

(B) 10 rad/s

(C) 12 rad/s

(D) 16 rad/s

3. The frequency response of a second-order system is given by

$$\frac{M}{\alpha} = \frac{X_2(j\omega)}{X_1(j\omega)}$$

The peak value of M, M_p, and the frequency at which it occurs, ω_r, are expressed in terms of the damping ratio, ζ, and natural frequency, ω_n.

$$M_p = \frac{1}{2\zeta\sqrt{1-\zeta^2}}$$

$$\omega_r = \omega_n\sqrt{1-2\zeta^2}$$

A polar plot of the system is shown.

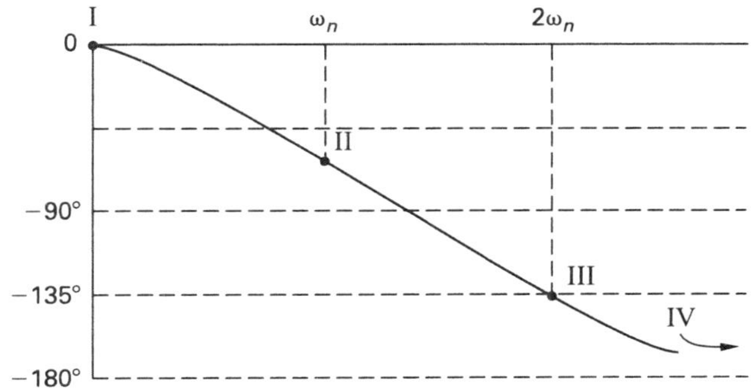

Where does the peak amplitude response occur?

(A) point II

(B) point IV

(C) between points I and II

(D) between points II and III

4. A block diagram is shown.

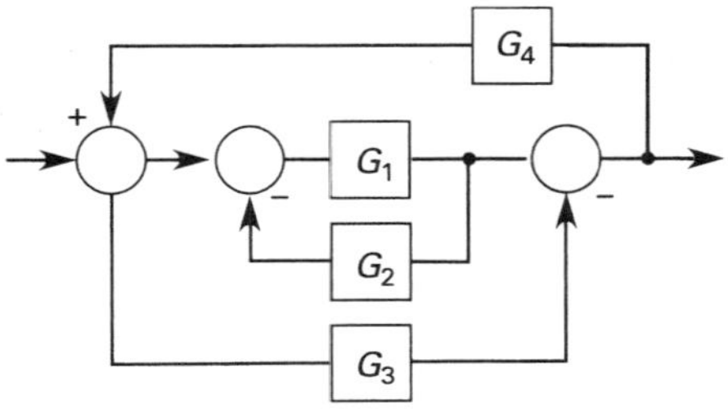

What is the overall system gain?

(A) $\dfrac{G_1+G_2+G_1G_2}{1+G_1G_3+G_1G_2+G_2G_3+G_1G_2G_3G_4}$

(B) $\dfrac{G_1+G_4+G_2G_3G_4}{1+G_1G_3+G_1G_2+G_2G_3+G_1G_2G_3G_4}$

(C) $\dfrac{G_1-G_2+G_3G_4}{1+G_1G_2+G_1G_4+G_3G_4+G_1G_2G_3G_4}$

(D) $\dfrac{G_1+G_3+G_1G_2G_3}{1+G_1G_2-G_1G_4-G_3G_4-G_1G_2G_3G_4}$

5. A block diagram is shown.

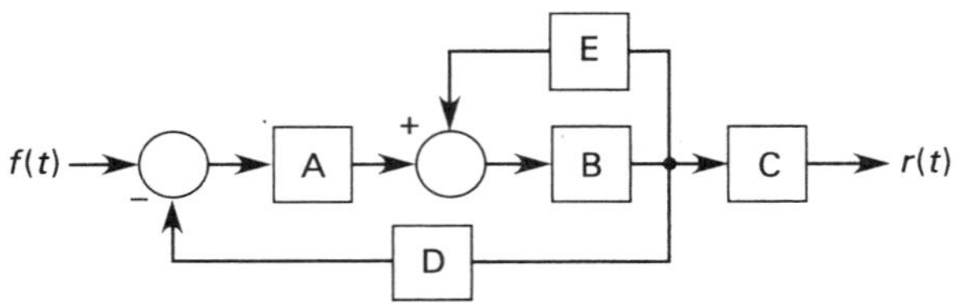

What is the overall system gain?

(A) $\dfrac{r(t)}{f(t)}=\dfrac{\text{ABE}}{1+\text{BC}-\text{ABE}}$

(B) $\dfrac{r(t)}{f(t)}=\dfrac{\text{ABE}}{1-\text{BC}+\text{ABE}}$

(C) $\dfrac{r(t)}{f(t)}=\dfrac{\text{ABC}}{1+\text{BE}-\text{ABD}}$

(D) $\dfrac{r(t)}{f(t)}=\dfrac{\text{ABC}}{1-\text{BE}+\text{ABD}}$

6. A block diagram is shown.

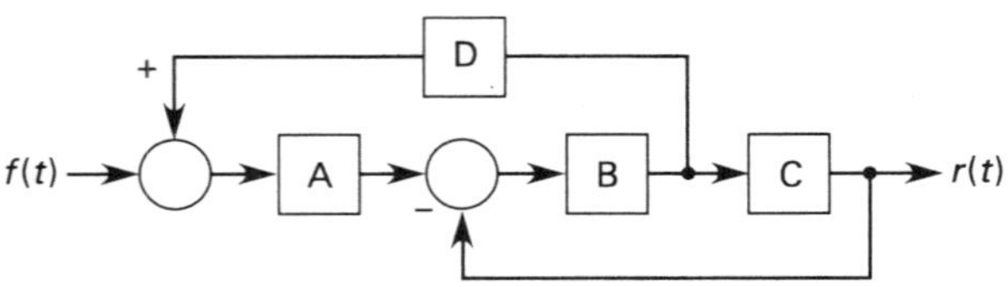

What is the overall system gain?

(A) $\dfrac{r(t)}{f(t)}=\dfrac{\text{ABD}}{1+\text{ABC}-\text{BD}}$

(B) $\dfrac{r(t)}{f(t)}=\dfrac{\text{ABD}}{1-\text{ABC}+\text{BD}}$

(C) $\dfrac{r(t)}{f(t)}=\dfrac{\text{ABC}}{1+\text{ABD}-\text{BC}}$

(D) $\dfrac{r(t)}{f(t)}=\dfrac{\text{ABC}}{1-\text{ABD}+\text{BC}}$

7. A block diagram is shown.

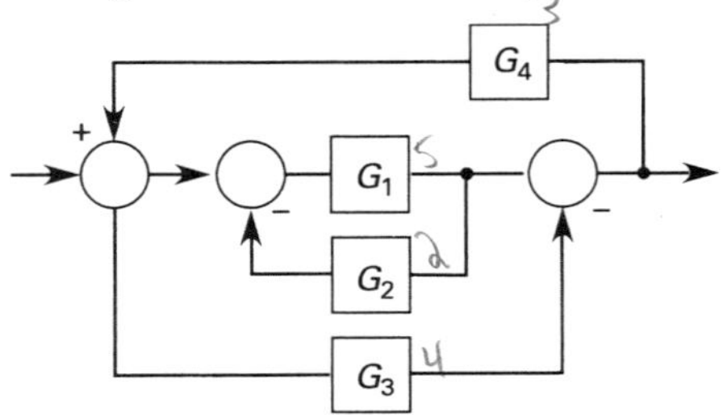

$G_1 = -5$ dB, $G_2 = 2$ dB, $G_3 = 4$ dB, and $G_4 = 3$ dB. What is the system sensitivity?

(A) $-3/34\ \text{dB}^{-1}$

(B) $-3/28\ \text{dB}^{-1}$

(C) $3/44\ \text{dB}^{-1}$

(D) $9/31\ \text{dB}^{-1}$

8. A control system with negative feedback is shown.

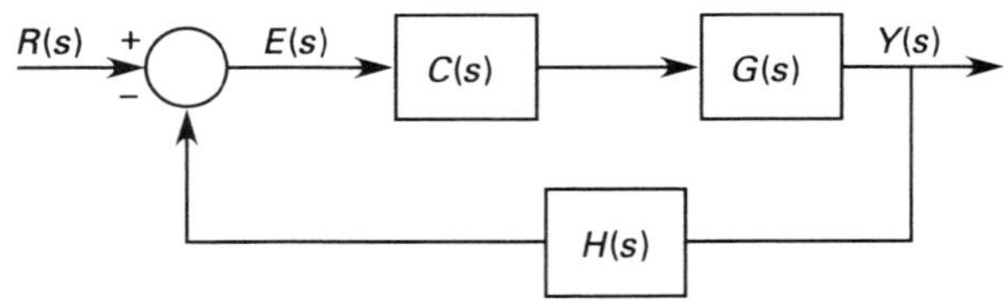

What is $E(s)$?

(A) $E(s) = R(s) - Y(s)$

(B) $E(s) = C(s)G(s)R(s)$

(C) $E(s) = \dfrac{G(s)R(s)}{1+C(s)H(s)}$

(D) $E(s) = \dfrac{R(s)}{1+C(s)G(s)H(s)}$

9. A control system with negative feedback is shown.

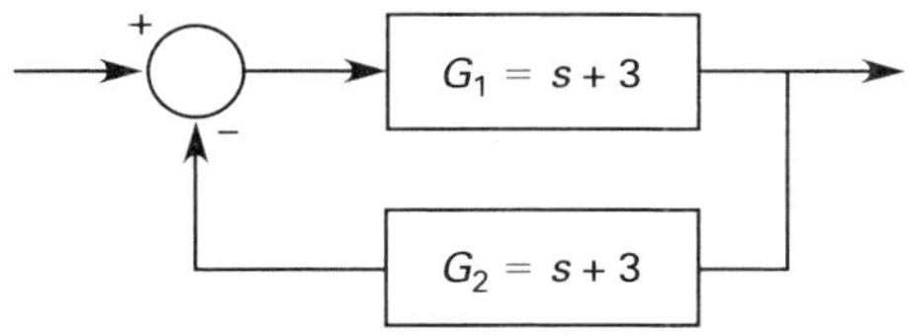

What is the time-domain transfer function of the system?

(A) $e^{-3t}\sin t$

(B) $e^{-3t}\cos t$

(C) $\dfrac{t+3}{1+(t+3)^2}$

(D) $\dfrac{1}{1+(t+3)^2}$

10. The transfer function of a control system is

$$T(s) = \frac{b_0 s^p + b_1 s^{p-1} + \cdots + b_p}{s^n + a_1 s^{n-1} + \cdots + a_n} \quad [a_n \neq 0]$$

What is the steady-state response to a step input for the control system?

(A) b_p/a_n

(B) b_p^2/a_n

(C) b_p/a_n^2

(D) b_p

11. A control system with negative feedback is constructed from linear time-invariant elements as shown.

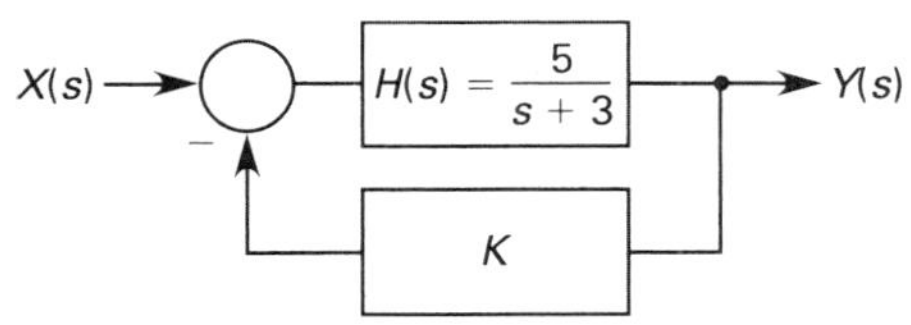

What is the requirement for the constant K such that the closed-loop system is stable?

(A) $K \leq -5/3$

(B) $K \leq -3/5$

(C) $K \geq -3/5$

(D) $K \geq 0$

SOLUTIONS

1. As ω increases to ω_r, M should increase to its peak value. The shapes of curve 2 and curve 4 are both correct. However, since M_p is always positive, only curve 2 is correct.

The answer is (B).

2. The general characteristic equation is

$$s^2 + 2\zeta\omega_n s + \omega_n^2 = 0$$

The characteristic equation for this system is

$$s^2 + 2s + 16 = 0$$

Find ω_n and ζ.

$$\omega_n = \sqrt{16} = 4$$

$$\begin{aligned}\zeta &= \frac{2s}{2\omega_n s} = \frac{1}{\omega_n} = \frac{1}{4}\\ &= 0.25\end{aligned}$$

The damped natural frequency of oscillation is

$$\begin{aligned}\omega_d &= \omega_n\sqrt{1-\zeta^2}\\ &= 4\sqrt{1-(0.25)^2}\\ &= 3.87 \quad (3.9\ \text{rad/s})\end{aligned}$$

The answer is (A).

3. As stated in the problem statement, the peak occurs at ω_r. From the equation for ω_r, when $\zeta = 0$, the peak occurs at $\omega_r = \omega_n$. When $\zeta > 0$, the peak occurs at $\omega_r < \omega_n$ (i.e., between points I and II).

The answer is (C).

4. Simplify the block diagram.

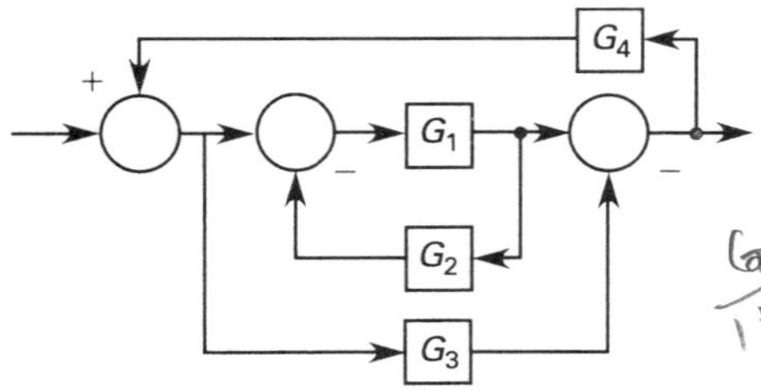

step 1

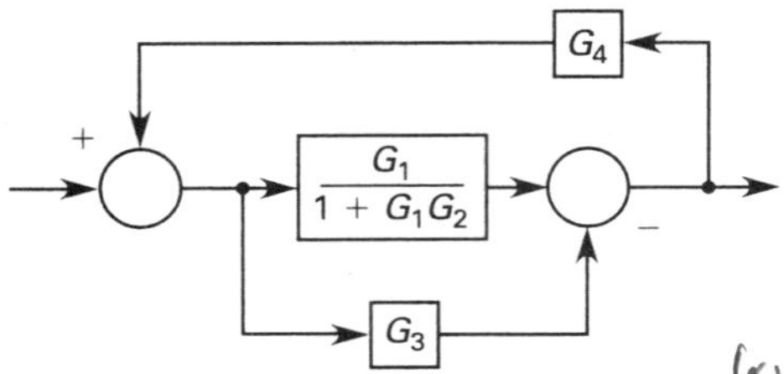

step 2

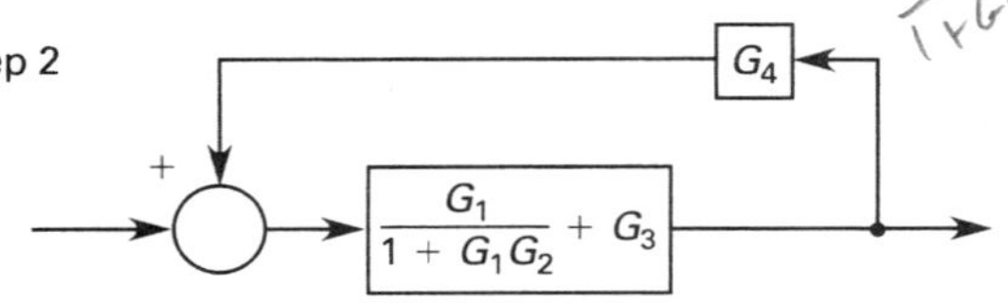

step 3

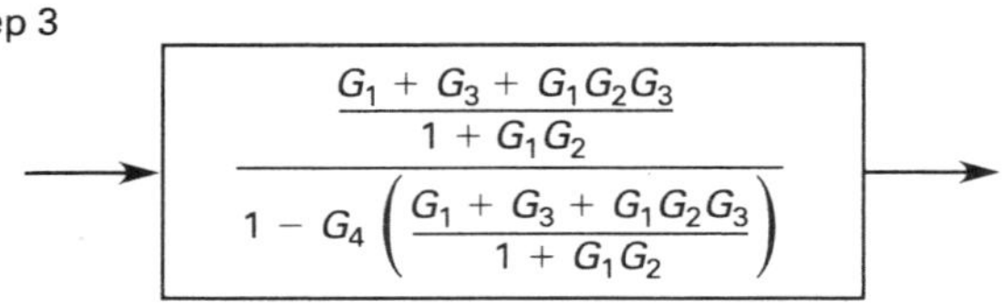

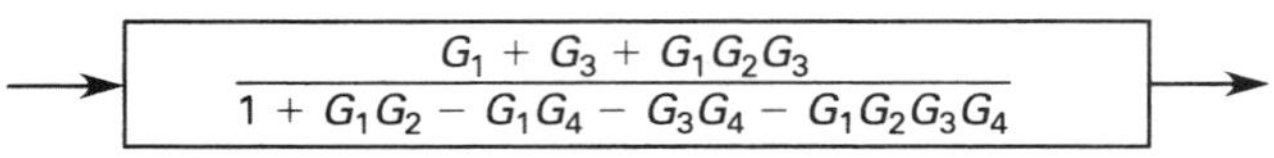

The overall system gain is

$$\frac{G_1 + G_3 + G_1G_2G_3}{1 + G_1G_2 - G_1G_4 - G_3G_4 - G_1G_2G_3G_4}$$

The answer is (D).

5. Simplify the block diagram.

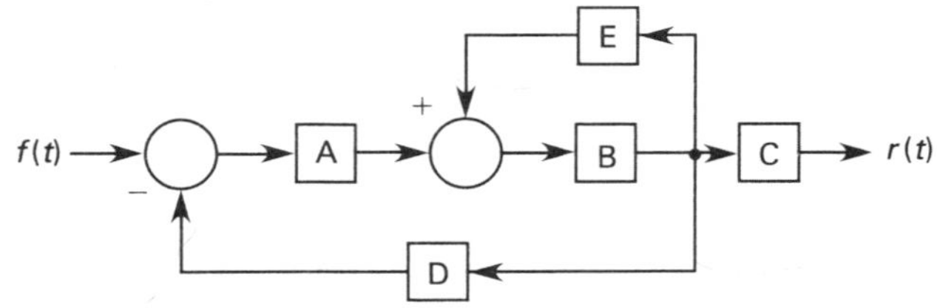

step 1

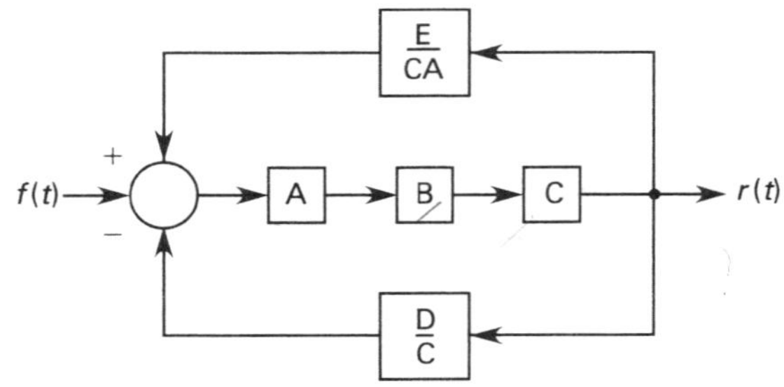

step 2

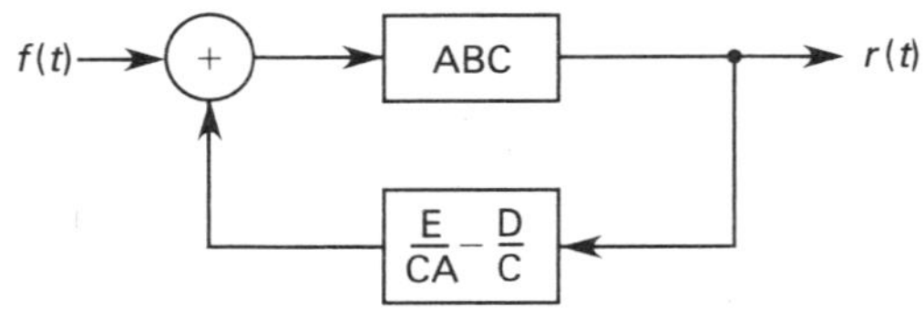

step 3

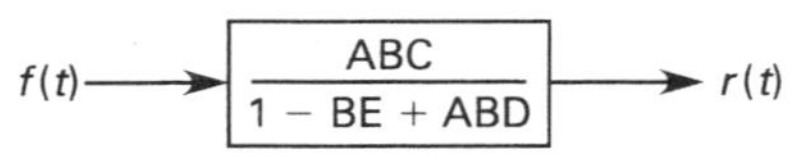

The overall system gain is

$$\frac{r(t)}{f(t)} = \frac{\mathrm{ABC}}{1 - \mathrm{BE} + \mathrm{ABD}}$$

The answer is (D).

6. Simplify the block diagram.

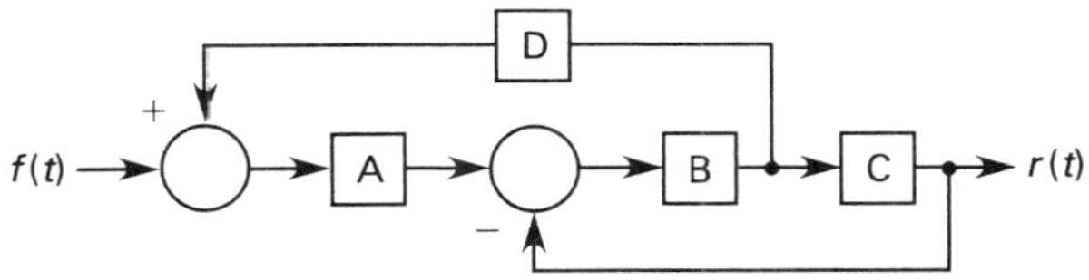

step 1

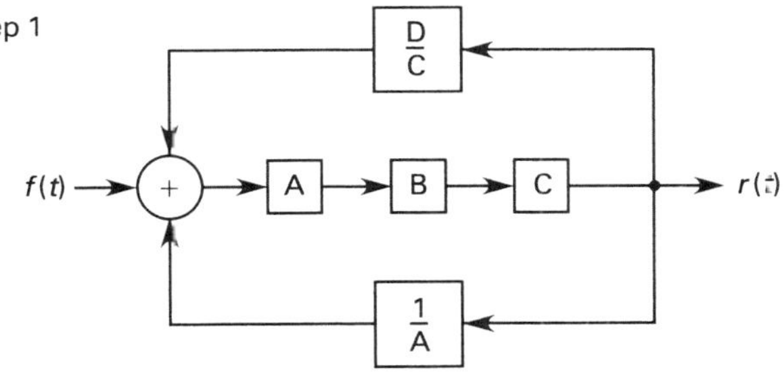

step 2

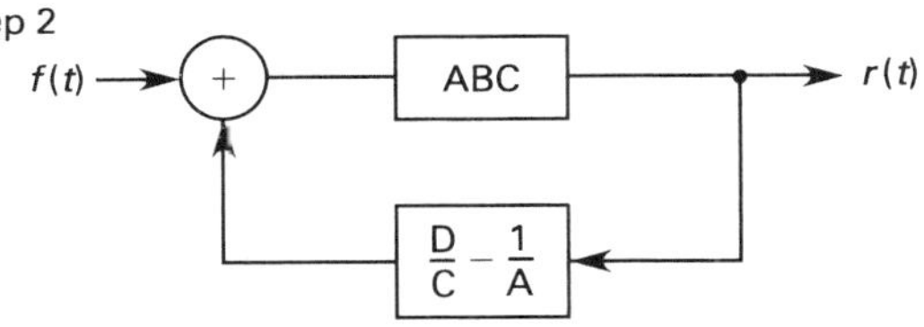

step 3

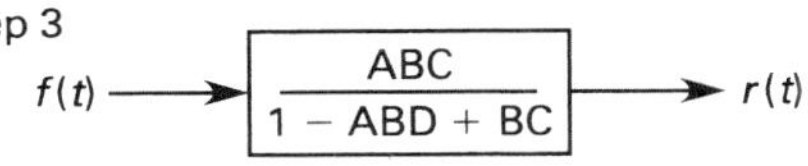

The overall system gain is

$$\frac{r(t)}{f(t)} = \frac{\text{ABC}}{1 - \text{ABD} + \text{BC}}$$

The answer is (D).

7. Simplify the block diagram.

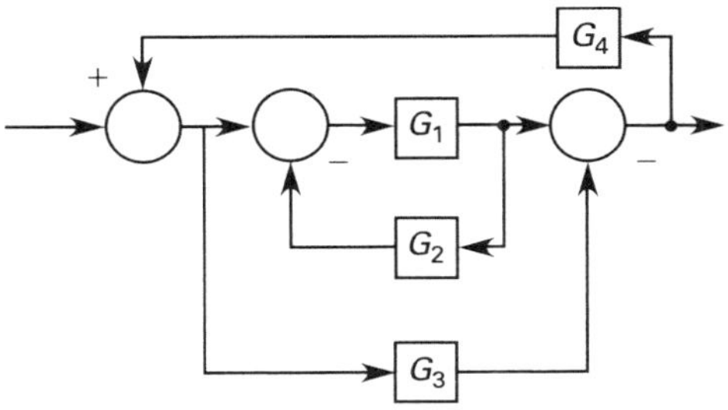

step 1

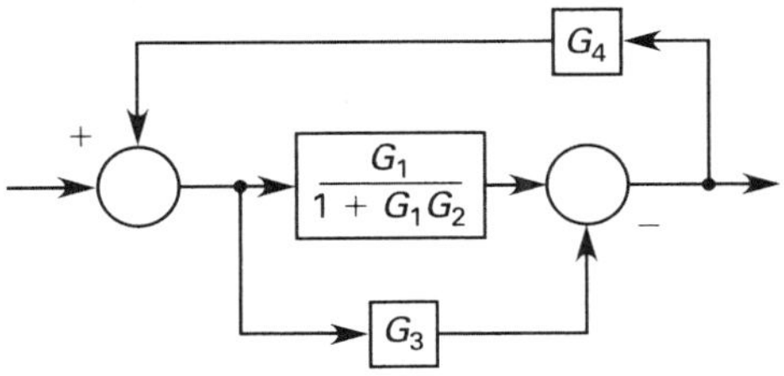

step 2

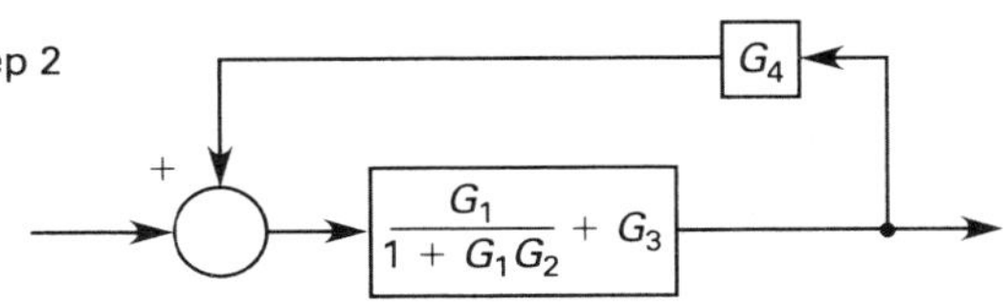

From step 2 of the simplified block diagram,

$$\begin{aligned} G(s) &= \frac{G_1}{1 + G_1 G_2} + G_3 \\ &= \frac{-5\text{ dB}}{1 + (-5\text{ dB})(2\text{ dB})} + 4\text{ dB} \\ &= 41/9\text{ dB} \\ H(s) &= G_4 \\ &= 3\text{ dB} \end{aligned}$$

$$\begin{aligned} G(s)H(s) &= \left(\frac{41}{9}\text{ dB}\right)(3\text{ dB}) \\ &= 41/3\text{ dB} > 0 \end{aligned}$$

The system has negative feedback. The sensitivity of the system is

$$\begin{aligned} S &= \frac{1}{1 + G(s)H(s)} \\ &= \frac{1}{1 + \dfrac{41}{3}\text{ dB}} \\ &= 3/44\text{ dB}^{-1} \end{aligned}$$

The answer is (C).

8. The error ratio for a closed-loop, negative feedback system with two forward-loop controllers in series is defined as

$$\frac{E(s)}{R(s)} = \frac{1}{1 + C(s)G(s)H(s)}$$

Therefore,

$$E(s) = \frac{R(s)}{1 + C(s)G(s)H(s)}$$

The answer is (D).

9. The control system circuit reduces to

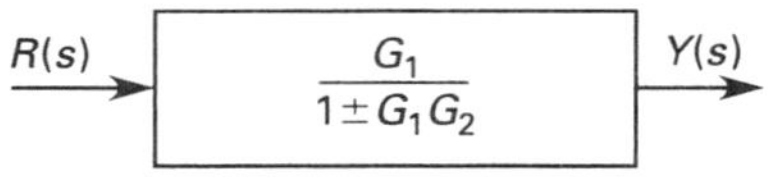

The problem statement specifies this is a negative feedback loop, so

$$\frac{Y(s)}{R(s)} = \frac{s+3}{1+(s+3)^2}$$

The inverse Laplace transform is

$$\frac{C(t)}{R(t)} = \mathcal{L}^{-1}\left(\frac{s+3}{1+(s+3)^2}\right) = e^{-3t}\cos t$$

The answer is (B).

10. Using the final value theorem, obtain the steady-state step response by substituting 0 for s in the transfer function.

$$Y(s) = \lim_{s\to 0} T(s) = T(0) = b_p/a_n$$

The answer is (A).

11. The system transfer function, $G(s)$, for the closed-loop system is

$$\begin{aligned} G(s) &= \frac{H(s)}{1+KH(s)} \\ &= \frac{\dfrac{5}{s+3}}{1+K\left(\dfrac{5}{s+3}\right)} \\ &= \frac{5}{s+3+5K} \end{aligned}$$

In order for the system to be stable, the pole of the closed-loop system transfer function needs to be in the left-hand side of the plane (i.e., must be negative).

$$\begin{aligned} s_1 &= -3 - 5K \le 0 \\ K &\ge -3/5 \end{aligned}$$

The answer is (C).

34 Signal Theory and Processing

PRACTICE PROBLEMS

1. An AM radio station broadcasts at 30 kW and 85% modulation. Most nearly, the power of the sidebands is

(A) 8 kW

(B) 10 kW

(C) 20 kW

(D) 30 kW

2. In a given AM environment, the highest frequency in the modulating signal is 1000 Hz. The frequency of the carrier is 100 kHz. Most nearly, the highest frequency in the AM signal is

(A) 49 kHz

(B) 51 kHz

(C) 99 kHz

(D) 100 kHz

3. What type of filter is shown?

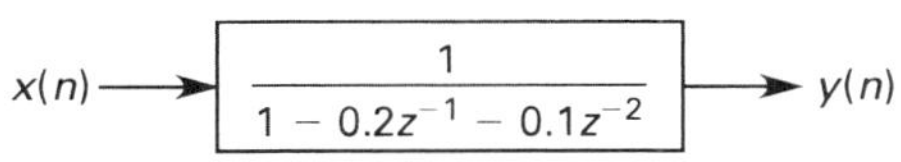

(A) finite impulse response

(B) infinite impulse response

(C) first order digital

(D) noncausal

4. An AM signal has a carrier wave that is given by $6 \cos 30{,}000t$. The modulation baseband signal that is added directly to the carrier amplitude is described by $\pi \cos 15{,}000t$. What will result from trying to recover the original signal?

(A) no overmodulation in the recovered signal

(B) aliasing in the recovered signal

(C) overmodulation in the recovered signal

(D) zero recovered signal

5. An angle modulated signal has a carrier signal of $x(t) = 10 \cos 10{,}000t$. The modulation baseband signal has an amplitude of 1 V, an angular frequency of 5000 rad/s, and a phase sensitivity, k_P, of 0.5 rad/V. Most nearly, the phase of the signal at $t = 5$ s is

(A) 1.2 rad

(B) 3.3 rad

(C) 5.0 rad

(D) 5.4 rad

6. In an amplitude modulated signal, a 900 kHz carrier is modulated by a music signal that has frequency components from 1 kHz to 10 kHz. The range of the frequencies generated for the upper sidebands is

(A) 440 kHz to 453 kHz

(B) 890 kHz to 899 kHz

(C) 899 kHz to 910 kHz

(D) 901 kHz to 910 kHz

7. Which figure represents the input step function $y = u[t-2]$?

(A)

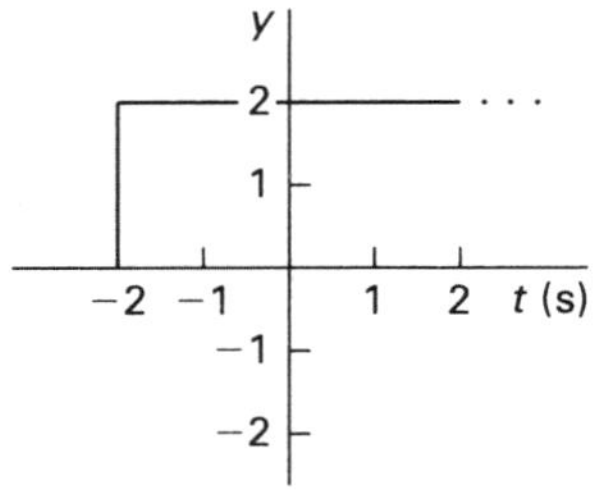

(B)

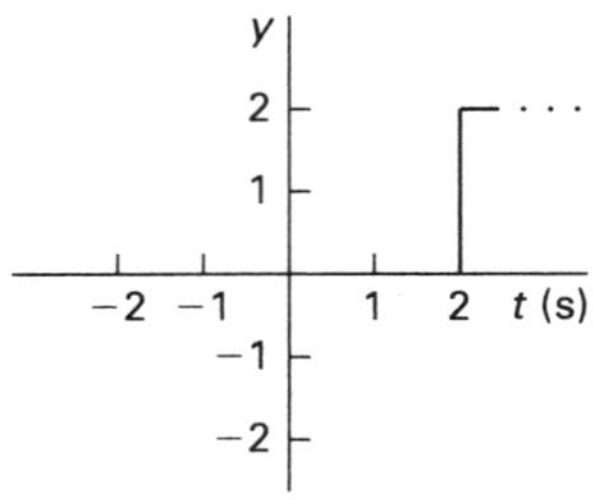

(C)

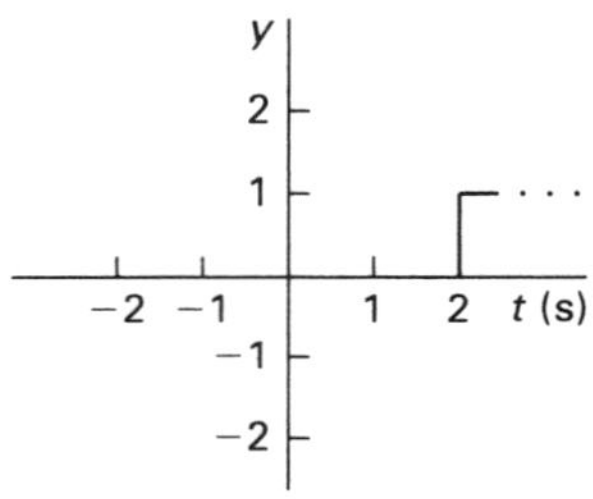

(D)

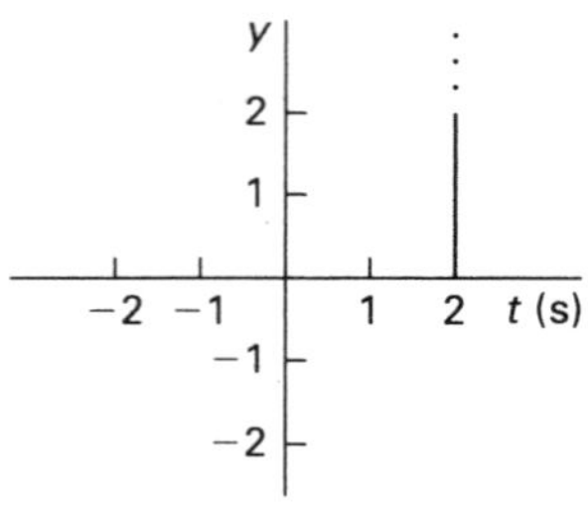

8. Which figure represents the input function $2\Pi(t/6)$?

(A)

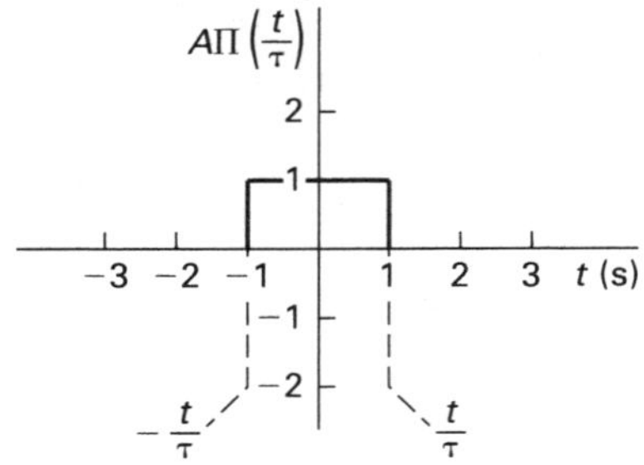

(B)

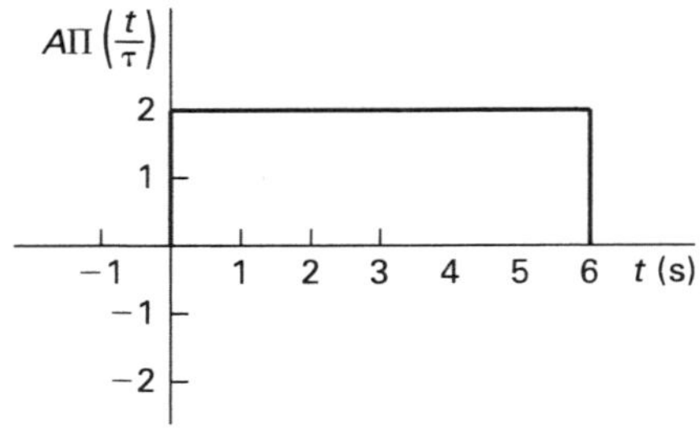

(C)

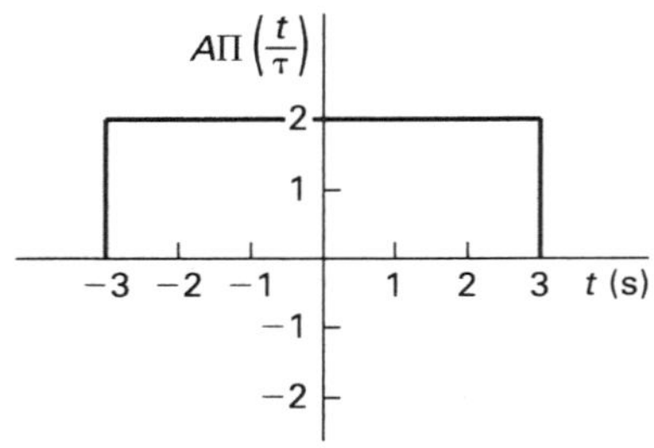

(D)

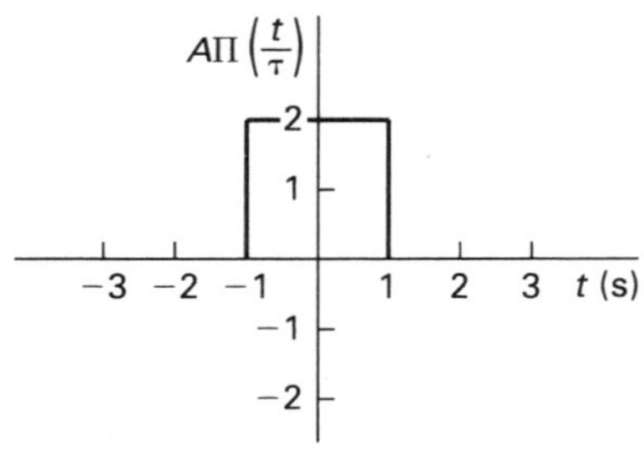

9. What is the Fourier transform of the triangular function $\Lambda(t/\tau)$?

(A) $\dfrac{1}{\tau}\text{sinc}\dfrac{\omega\tau}{2}$

(B) $\dfrac{1}{\sqrt{3}}\text{sinc}^2\dfrac{\tau}{2}$

(C) $\text{sinc}^2\tau f$

(D) $\tau\text{sinc}^2\dfrac{\tau\omega}{2\pi}$

10. Consider the following complex Fourier series representation of a signal.

$$\sum_{n=-\infty}^{\infty} j\left(\frac{(-1)^n}{n}\right)e^{jn2\pi f_0 t}$$

Most nearly, what is the average power of the third harmonic of the signal?

(A) 1/9

(B) 1/6

(C) 2/9

(D) 1/3

11. Consider a normal amplitude modulated signal in the frequency domain. Which portion of the signal carries the information?

(A) carrier frequency

(B) lower sideband only

(C) upper sideband only

(D) both the lower and upper sidebands

12. In digital communication, which of the following is analogous to angle modulation in FM communication?

(A) DSB-LC

(B) FSK

(C) PSK

(D) polar modulation

13. A modulated broadband signal has a modulation index, m_{FM}, of 1.5 and a modulation frequency of 5 kHz. Most nearly, what is the maximum frequency deviation?

(A) 2.5 kHz

(B) 3.3 kHz

(C) 3.5 kHz

(D) 7.5 kHz

14. A narrowband FM signal has a message bandwidth of 5 kHz. Most nearly, what is the 98% power bandwidth of the signal?

(A) 2.5 kHz

(B) 5.0 kHz

(C) 7.5 kHz

(D) 10 kHz

15. The frequency deviation ratio of a wideband FM signal is 20, and the modulating frequency is 7 kHz. Most nearly, the bandwidth is

(A) 3 kHz

(B) 30 kHz

(C) 140 kHz

(D) 300 kHz

16. A pulse-code modulation (PCM) signal is transmitted with a message frequency of 20 kHz and a quantization level of 16. If the quantization level is increased to 32, the minimum required bandwidth will increase by a factor of

(A) 1.25

(B) $\sqrt{2}$

(C) $\sqrt{3}$

(D) 4

Comm./Signal Processing

SOLUTIONS

1. Use the equation for total transmitted power and solve for the carrier power.

$$P_{\text{total}} = P_{\text{carrier}}\left(1+\frac{m^2}{2}\right)$$

$$P_{\text{carrier}} = \frac{P_{\text{total}}}{1+\frac{m^2}{2}} = \frac{30\text{ kW}}{1+\frac{(0.85)^2}{2}} = 22.04\text{ kW}$$

P_{total} is the total transmitted power of the sidebands and carrier, so

$$\begin{aligned} P_{\text{total}} &= P_{\text{carrier}} + P_{\text{sideboard}} \\ P_{\text{sideboard}} &= P_{\text{total}} - P_{\text{carrier}} = 30\text{ kW} - 22.04\text{ kW} \\ &= 7.96\text{ kW} \quad (8\text{ kW}) \end{aligned}$$

The answer is (A).

2. In an AM environment, the frequencies of the modulating signal are shifted by the carrier frequency. The resulting signal will have its frequencies concentrated around the carrier frequency.

Therefore, the highest frequency in the AM signal will be the sum of the carrier frequency and the highest frequency in the modulating signal.

$$100\text{ kHz} + \frac{1000\text{ Hz}}{1000\,\frac{\text{Hz}}{\text{kHz}}} = 101\text{ kHz} \quad (100\text{ kHz})$$

The answer is (D).

3. If $h(n)$ is the impulse response of the filter, then

$$\begin{aligned} Y(z) &= X(z)H(z) \\ &= X(z)\left(\frac{1}{1-0.2z^{-1}-0.1z^{-2}}\right) \\ X(z) &= Y(z)(1-0.2z^{-1}-0.1z^{-2}) \end{aligned}$$

The inverse z-transform of $z^{-k}Y(z)$ is $y(n-k)$ (shift property), so the inverse transform of the preceding equation is

$$x(n) = y(n) - 0.2y(n-1) - 0.1y(n-2)$$

Solving for $y(n)$,

$$y(n) = x(n) + 0.2y(n-1) + 0.1y(n-2)$$

The impulse response can be determined by letting $x(n) = \delta(n)$.

n	$x(n)$	$Y(n) = h(n)$
<0	0	0
0	1	1
1	0	0.2
2	0	0.14
3	0	0.048
4	0	0.024
5	0	0.0096
⋮	⋮	⋮

Although $h(n)$ decreases as n increases, it never reaches zero. The impulse response is infinite. The filter is an infinite impulse response (IIR) filter.

The answer is (B).

4. The AM signal is of the form

$$\begin{aligned} x_{\text{AM}}(t) &= A_c[A+m(t)]\cos(2\pi f_c t) \\ &= A'_c[1+am_n(t)]\cos(2\pi f_c t) \\ &= 6(1+\pi\cos 15{,}000t)\cos 30{,}000t \end{aligned}$$

The cosine has a maximum value of 1 and a minimum value of −1, so the maximum and minimum amplitudes of the modulating signal are

$$\begin{aligned} \max(x_{\text{AM}}(t)) &= 6\big(1+\max(m_n(t))\big)\big(\max(\cos 30{,}000t)\big) \\ &= (6)\big(1+\max(\pi\cos 15{,}000t)\big)(1) \\ &= (6)(1+\pi)(1) \\ &= 24.85 \\ \min(x_{\text{AM}}(t)) &= 6\big(1+\min(m_n(t))\big)\big(\max(\cos 30{,}000t)\big) \\ &= (6)\big(1+\min(\pi\cos 15{,}000t)\big)(1) \\ &= (6)(1-\pi)(1) \\ &= -12.85 \end{aligned}$$

Because $m_n(t)$ has values that are less than −1, there will be overmodulation. Overmodulation results in distortion when trying to recover the original signal.

The answer is (C).

5. The modulation baseband signal is

$$\begin{aligned} m(t) &= A\cos\omega_m t \\ &= (1\text{ V})\cos\left(5000\,\frac{\text{rad}}{\text{s}}\right)t \end{aligned}$$

The angle modulation at $t = 5$ s is

$$\begin{aligned}\phi(t) &= k_p m(t) \\ &= \left(0.5\ \frac{\text{rad}}{\text{V}}\right)(1\ \text{V})\cos\left(\left(5000\ \frac{\text{rad}}{\text{s}}\right)(5\ \text{s})\left(\frac{360°}{2\pi}\right)\right) \\ &= 0.35\ \text{rad}\end{aligned}$$

The general equation for an angle modulated signal is

$$v(t) = A_c\cos(\omega_c t + \phi(t))$$

The phase of the modulated signal at $t = 5$ s is

$$\begin{aligned}\omega_c t + \varphi(t) &= \left(10{,}000\ \frac{\text{rad}}{\text{s}}\right)(5\ \text{s}) + 0.35\ \text{rad} \\ &= 50{,}000.35\ \text{rad}\end{aligned}$$

Subtract the whole cycles, then convert the remaining fraction of a cycle to radians.

$$\frac{50{,}000.35\ \text{rad}}{2\pi\ \frac{\text{rad}}{\text{cycle}}} = 7957.803\ \text{cycles} \quad (0.803\ \text{cycle})$$

$$(0.803\ \text{cycle})\left(2\pi\ \frac{\text{rad}}{\text{cycle}}\right) = 5.045\ \text{rad} \quad (5.0\ \text{rad})$$

The answer is (C).

6. The highest frequency of the modulating signal is $f_h = 10$ kHz. The lowest frequency of the modulating signal is $f_l = 1$ kHz. The carrier frequency is $f_c = 900$ kHz. The upper sidebands will include the frequencies from $f_c + f_l$ to $f_c + f_h$. The range of the upper sidebands is from 901 kHz to 910 kHz.

The answer is (D).

7. The function $y = u[t - 2]$ steps to value 1 at $t = 2$.

The answer is (C).

8. The input function $2\Pi(t/6)$ has an amplitude of 2 and a pulse width of 6 s and is symmetrical around the vertical axis. Only option C shows all these features.

Option A has an amplitude of 1 and a pulse width of 1 s. Option B has the right amplitude and pulse width, but is not symmetrical about the vertical axis. Option D has the right amplitude but a pulse width of 2 s.

The answer is (C).

9. The triangular function $\Lambda(t/\tau)$ starts at $t = -\tau$, increases linearly to a value of one at $t = 0$, then decreases linearly to a value of zero at $t = \tau$.

From a table of Fourier transform pairs, the Fourier transform of the triangular function is

$$X(f) = \tau\text{sinc}^2\tau f$$

As $\omega = 2\pi f$,

$$X(\omega) = \tau\text{sinc}^2\frac{\tau\omega}{2\pi}$$

The answer is (D).

10. The coefficients of the exponential terms, X_n, represent the voltage and are

$$j\left(\frac{(-1)^n}{n}\right)$$

The magnitude of the coefficients is

$$\left|j\left(\frac{(-1)^n}{n}\right)\right| = \frac{(-1)^n}{n}$$

Squaring the results (to represent power) gives

$$\left(\frac{(-1)^n}{n}\right)^2$$

The average power in the DC component and the first N harmonics is

$$P = X_0^2 + 2\sum_{n=0}^{N}|X_n|^2$$

The value of X_0 is 0. The power of just the third harmonic alone is

$$\begin{aligned}P_3 = 2|X_3|^2 &= 2\left(\frac{(-1)^n}{n}\right)^2 = (2)\left(\frac{(-1)^3}{3}\right)^2 \\ &= \frac{2}{9}\end{aligned}$$

This indicates that 2/9 of the harmonic power lies with the third harmonic.

The answer is (C).

11. In the frequency domain, the AM signal's power is distributed between the carrier frequency and two sidebands. Both sidebands carry identical information, so either can be used to obtain the original information.

The answer is (D).

12. Analog frequency modulation (FM) developed first. Following that, digital communication developed the concept of frequency-shift keying (FSK).

The other options listed are double-sideband, large carrier (DSB-LC, a type of AM communication), phase-shift keying (PSK), and polar modulation (which is analogous to quadrature modulation).

The answer is (B).

13. Use the definition of the modulation index, and rearrange to solve for the maximum frequency deviation.

$$m_{\text{FM}} = \frac{\Delta\omega}{\omega_{\text{mod}}}$$

$$\begin{aligned}\Delta\omega &= m_{\text{FM}}\omega_{\text{mod}} = (1.5)(5\ \text{kHz})\\ &= 7.5\ \text{kHz}\end{aligned}$$

The modulation index is greater than one, so this is a broadband signal.

The answer is (D).

14. A narrowband FM signal has only a single pair of sidebands.

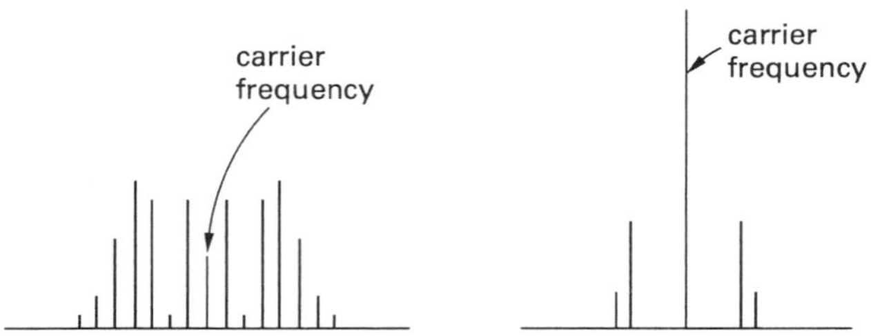

(a) wideband frequency spectrum

(b) narrowband frequency spectrum

The 98% power bandwidth for a narrowband signal is given by

$$B \cong 2W = (2)(5\,\text{kHz}) = 10\,\text{kHz}$$

The answer is (D).

15. The modulating frequency of 7 kHz is the message bandwidth, W. The bandwidth is

$$\begin{aligned}B &\cong 2(D+1)W\\ &= (2)(20+1)(7\,\text{kHz})\\ &= 294\,\text{kHz}\quad (300\,\text{kHz})\end{aligned}$$

The answer is (D).

16. The minimum required bandwidth, B, is proportional to

$$B \propto 2nW = 2W\log_2 q$$

If q is increased from 16 to 32, the value of $\log_2 q$ increases from 4 to 5, and B will increase by a factor of 5/4, or 1.25.

The answer is (A).

35 Computer Hardware and Fundamentals

PRACTICE PROBLEMS

1. In virtual memory systems, large programs

(A) are executed in memory

(B) can be executed using less memory space than in RAM

(C) are paged into and out of unused video memory

(D) run faster due to fewer disk accesses

2. ASCII characters are sent over a 64 kb/s communications channel. One bit per character is used for parity checking. Most nearly, how many ASCII characters can be transmitted over the channel in 5 s?

(A) 35 000 characters

(B) 40 000 characters

(C) 42 000 characters

(D) 46 000 characters

3. A simple controller board has two thousand 8-bit memory locations and two 8-bit registers. How many different states can this board be in?

(A) 2002

(B) $(2)^8$

(C) $(2)^{2002}$

(D) $(2)^{16,016}$

4. What information must be specified in the instruction for a direct memory access process?

(A) the address of the address of the operand

(B) the address of the operand

(C) the operand itself

(D) the register and base addresses of the operand

5. The acronym RDRAM refers to

(A) Rambus dynamic random access memory

(B) Rambus direct random access memory

(C) redundant dual random access memory

(D) random duplex random access memory

6. Flash memory is classified as what type of memory?

(A) ROM

(B) WORM

(C) RAM

(D) EPROM

7. What type of memory is lost when power to a device is removed?

(A) RAM

(B) ROM

(C) PROM

(D) EPROM

8. Which of the following is the best description of a nibble?

(A) half a byte of memory

(B) a read-only operation

(C) a CPU operation that is performed in half a clock cycle

(D) a CPU operation that is performed in one clock cycle

9. Which of the following is the best description of a bit?

(A) a basic unit of a computer used to encode a single character of text

(B) the smallest portion of computer memory that can represent a distinct computer address

(C) a binary digit that can have one of only two possible values

(D) computer memory that can represent either of two possible states

10. Which of the following is the best description of a buffer?

(A) a region where extra information goes once the main memory is full

(B) a temporary storage region

(C) the main memory

(D) a permanent memory region where start-up information is stored

11. How does a central processing unit (CPU) determine whether it is executing instructions from a commercial database management program or from a program executed by an online programmer?

(A) The micro-operations used in the programs are different.

(B) One kind of program uses memory references, and the other uses register references.

(C) One kind of program uses compiled code, and the other uses interpreted code.

(D) The CPU does not know where instructions originate.

12. ASCII coding accommodates a standard set of English characters for data transfer. There are several extended sets of ASCII characters that are dependent on the operating system. How many bits represent the standard and extended sets, respectively?

(A) 7 bits, 7 bits

(B) 7 bits, 8 bits

(C) 8 bits, 12 bits

(D) 8 bits, 16 bits

SOLUTIONS

1. In a virtual memory system, large programs are systematically paged into and out of disk space and can be executed using a smaller memory space.

The answer is (B).

2. ASCII code requires a minimum of 7 bits per character. With 1 bit for parity checking, each character requires 8 bits to transmit. The number of ASCII characters that can be transmitted over a 64 kb/s channel in 5 s is

$$\begin{aligned}\text{characters} &= \frac{\left(\left(64\ \frac{\text{kb}}{\text{s}}\right)\left(1000\ \frac{\text{b}}{\text{kb}}\right)\right)(5.0\ \text{s})}{8\ \frac{\text{bits}}{\text{character}}} \\ &= 40\,000\ \text{characters}\end{aligned}$$

The answer is (B).

3. The number of bits in the memory and registers is

$$(8)(2000 + 2) = 16{,}016$$

Each bit can take on two different values. The number of different states is

$$(2)^{16{,}016}$$

The answer is (D).

4. In direct memory access, the instruction must specify the address of the operand.

The answer is (B).

5. RDRAM is an acronym for Rambus dynamic random access memory. RDRAM computer chips use Rambus signaling level technology to achieve high random-access speeds.

The answer is (A).

6. Flash memory is a type of erasable programmable read-only memory (EPROM).

The answer is (D).

7. ROM (read-only memory), PROM (programmable read-only memory), EPROM (erasable programmable read-only memory), and WORM (write once, read many) retain their information when power to the device is removed. RAM (random-access memory) does not retain information when the power is removed.

The answer is (A).

8. A nibble is a block of four contiguous bits (i.e., half a byte) of memory.

The answer is (A).

9. A bit is a binary digit, a basic unit of digital information. A bit can have one of only two possible values, which are usually represented as 0 and 1.

The answer is (C).

10. A buffer is a temporary storage region that holds data until it is used.

The answer is (B).

11. A central processing unit cannot determine the source of instructions. The CPU will execute instructions regardless of their origins.

The answer is (D).

12. The standard set of ASCII characters uses 7 bits in 128 combinations, each combination representing one of the English characters. In the extended set, 8 bits are used to encode 256 characters.

The answer is (B).

36 Networking Systems

PRACTICE PROBLEMS

1. Which of the following best describes a relational database?

(A) a collection of data organized into tables whose column order determines the relationships between the data

(B) a collection of data organized into independent tables whose data can be reorganized by software without changing the original tables

(C) the same as a network database (with no duplication of data in tables)

(D) a collection of data organized in a treelike structure

2. Which of these statements about local area networks (LANs) is true?

(A) Most devices in a LAN are connected in master-slave fashion.

(B) Internet service providers (ISPs) provide LAN services to customers.

(C) There is a single administrative system in a LAN.

(D) A LAN may be distributed over a large geographical area using telecommunications.

3. Which of these is a method of network communication between a source node and a destination node along a dedicated path that is not shared with any other communication?

(A) packet switching

(B) message switching

(C) interface protocol

(D) circuit switching

4. These four diagrams represent network topologies. Which is a star topology?

(A)

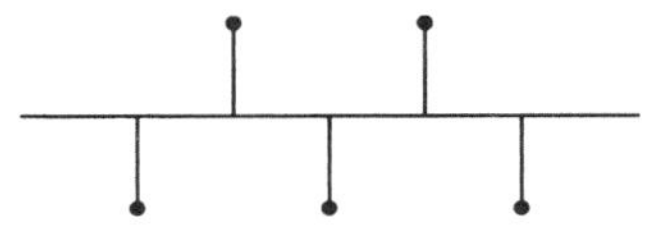

(B)

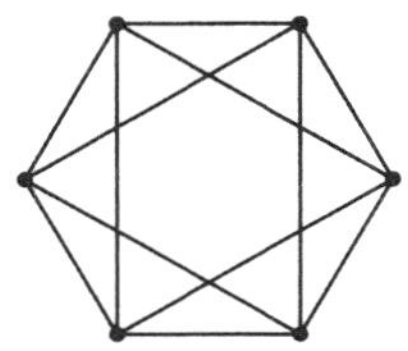

(C)

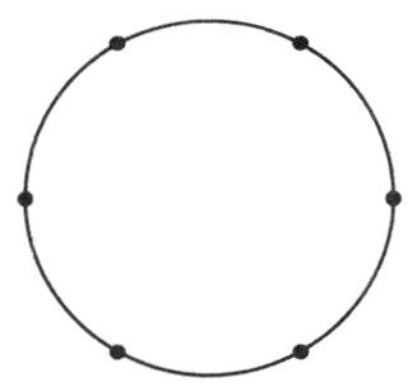

(D)

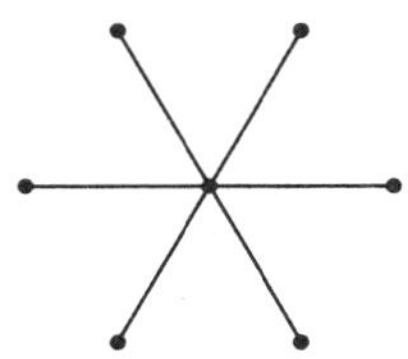

5. In which kind of local area network (LAN) topology will the failure of any single host cause the failure of the entire network?

(A) bus

(B) ring

(C) star

(D) mesh

SOLUTIONS

1. A database is a structured collection of records or data. A computer database relies on software to organize the storage of data. The software models the database structure in what are known as *database models.*

The database model in most common use today is the *relational model.* A *relational database* is a collection of data items organized as a set of formally described tables. Data from these tables can be accessed or reassembled in many different ways without having to reorganize the tables themselves.

In a relational database, data is fitted into predefined categories. Each table (which is sometimes called a *relation*) contains one or more data categories in columns. Each row contains a unique instance of data for the categories defined by the columns.

For example, a typical business order entry database would include a table that described a customer with columns for name, address, phone number, and so forth. Another table would describe an order: product, customer, date, sales price, and so forth. Each user of the database can obtain a view of the database that fits that user's needs. For example, a branch office manager might want a report on all customers who bought a certain product after a certain date, while a financial services manager could, from the same tables, obtain a report on accounts that needed to be paid.

Other models, such as the hierarchical model and the network model, use a more explicit representation of relationships. In a *hierarchical model*, data is organized into a treelike structure, which implies a single upward link in each node to describe the nesting along with a sort field to keep the records in a particular order in each list on the same level. The *network model* tends to store records with links to other records. Associations are tracked via "pointers," which can be node numbers or disk addresses. Most network databases tend to include also some form of hierarchical model.

The answer is (B).

2. In a LAN, the communication of the networked computers is regulated by a single administrative system. Option C is true.

Devices in personal area networks (PANs) are connected in master-slave fashion; most devices in LANs are not. Option A is false.

ISP services connect LANs and single users to other resources through a metropolitan area network (MAN). Option B is false.

Wide area networks (WANs) are used to connect MANs and LANs over large geographical areas using telecommunications. Option D is false.

The answer is (C).

3. Circuit switching establishes a dedicated path for communication between a source node and a destination node. The circuit may be permanent, or it may be temporarily connected for transmission and then disconnected when communication is complete. Option D is correct.

In packet switching, a message from a source node is broken into packets. Header information is added to each packet so that the message can be reassembled at the destination node. Packets are transmitted to intermediate locations, where they are stored and forwarded toward their destination, possibly by way of further intermediate locations. The use of intermediate locations makes it possible for more than one communication to share the same path, so Option A is incorrect.

Message switching communication transmits a complete message to an intermediate location where the message is stored and forwarded toward its destination, possibly by way of further intermediate locations. The use of intermediate locations makes it possible for more than one communication to share the same path, so Option B is incorrect.

The phrase "interface protocol" is a generic term for a communication mechanism, but does not specify a method. Option C is incorrect.

The answer is (D).

4. In the star network topology, each of the network's clients is connected, using point-to-point communication, to a central routing device called a *hub.* The hub transfers messages from one device to another within the network. The diagram in Option D shows the hub as the dot in the center of the diagram, and each other device on the network is represented by a dot connected to the hub.

The answer is (D).

5. In a ring topology, each host is needed for messages to pass from one host to another. The failure of any single host will result in the failure of the entire network. Option B is correct.

In a bus topology, the failure of the communication line that connects the devices will cause the failure of the entire network, but failure of a single connection or a single client on the network will not cause the rest of the network to fail. Option A is incorrect.

In a star topology, the failure of the central routing device (called the hub) will cause the entire network to fail, but the failure of a single client or of the communication line to a client will cause the failure of communication only to that one client. Option C is incorrect.

Failure of a single host in the mesh topology will cause failure of communication only to that host; the rest of the network can function. Option D is incorrect.

The answer is (B).

37 Digital Logic

PRACTICE PROBLEMS

1. In the circuit shown, $X=1$, $Y=0$, and $Z=1$.

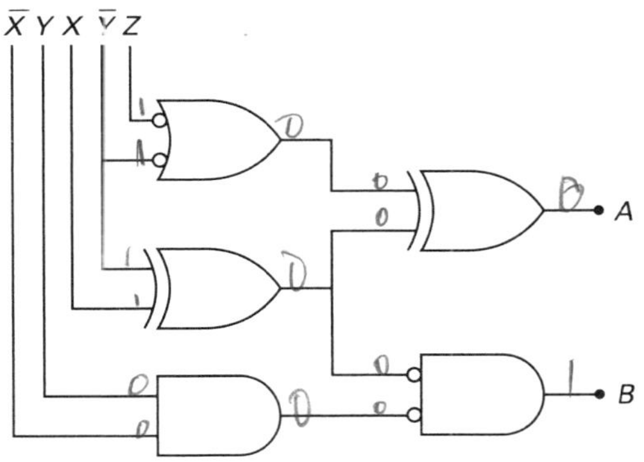

What are the outputs, A and B?

(A) $A=0$, $B=0$

(B) $A=0$, $B=1$

(C) $A=1$, $B=0$

(D) $A=1$, $B=1$

2. What is the output of the following logic circuit?

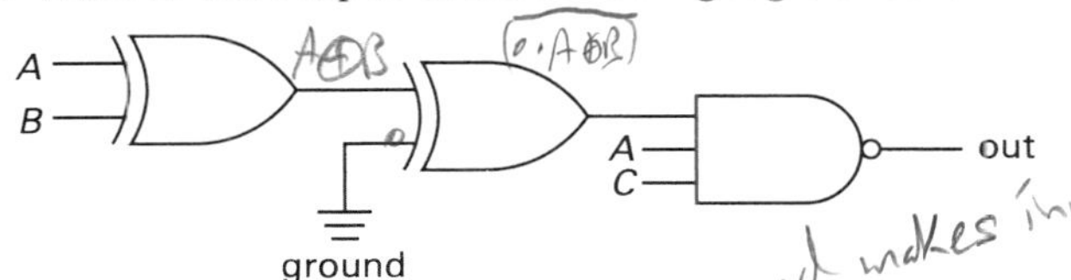

(A) $A \cdot B \cdot C$

(B) $A \cdot B \cdot C + A \cdot C$

(C) $A \cdot B \cdot C + \overline{A} \cdot B \cdot C + A \cdot \overline{B} \cdot C + A \cdot B \cdot \overline{C}$

(D) $A \cdot \overline{B} \cdot C$

3. Which column of the truth table corresponds to the logic circuit shown?

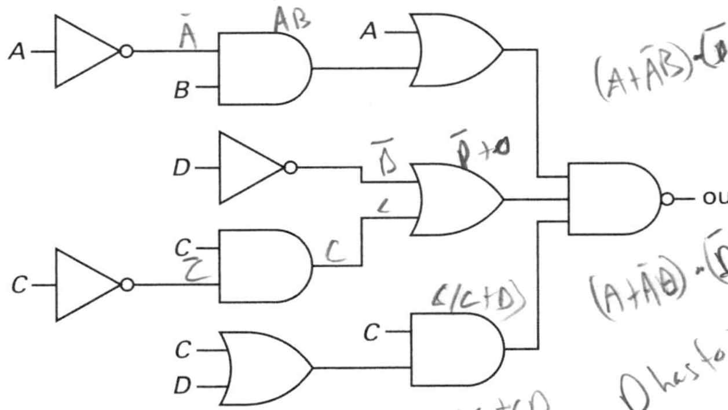

A	B	C	D	I	II	III	IV
0	0	0	0	1	0	0	1
0	0	0	1	1	0	0	0
0	0	1	0	0	0	0	1
0	0	1	1	0	0	0	0
0	1	0	0	0	0	0	1
0	1	0	1	0	0	0	0
0	1	1	0	1	1	1	1
0	1	1	1	0	0	0	0
1	0	0	0	0	0	0	1
1	0	0	1	0	0	0	0
1	0	1	0	1	1	1	1
1	0	1	1	0	0	0	0
1	1	0	0	0	1	0	1
1	1	0	1	0	0	0	0
1	1	1	0	1	1	1	1
1	1	1	1	0	1	0	0

(A) column I

(B) column II

(C) column III

(D) column IV

4. A logic circuit consists of an AND, NAND, and NOR gate, as shown.

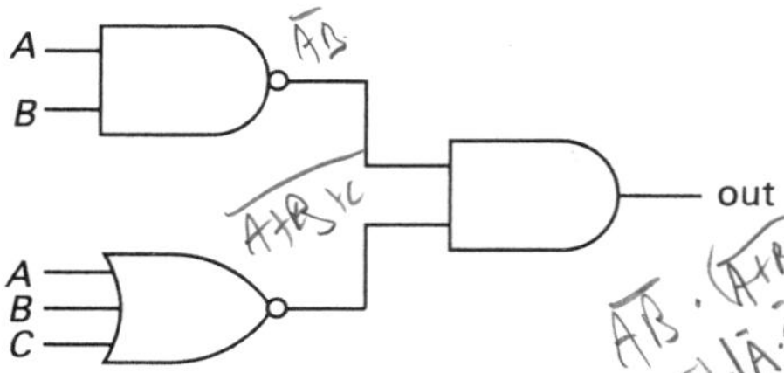

What is the output of the AND gate?

(A) 0

(B) $(A+B)\cdot(A\cdot B\cdot C)$

(C) $(\overline{A}\cdot\overline{C})+(\overline{B}\cdot\overline{C})$

(D) $\overline{A}\cdot\overline{B}\cdot\overline{C}$

5. A logic circuit consists of five gates as shown.

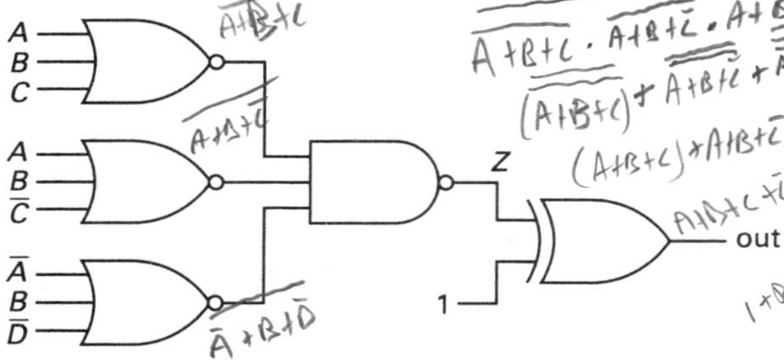

What is the output?

(A) 0

(B) 1

(C) $A\cdot B\cdot C+B+A\cdot\overline{C}\cdot\overline{D}$

(D) $\overline{A}\cdot\overline{B}\cdot\overline{C}+\overline{A}\cdot\overline{B}\cdot C+A\cdot\overline{B}\cdot D$

6. A logic circuit consists of four gates as shown.

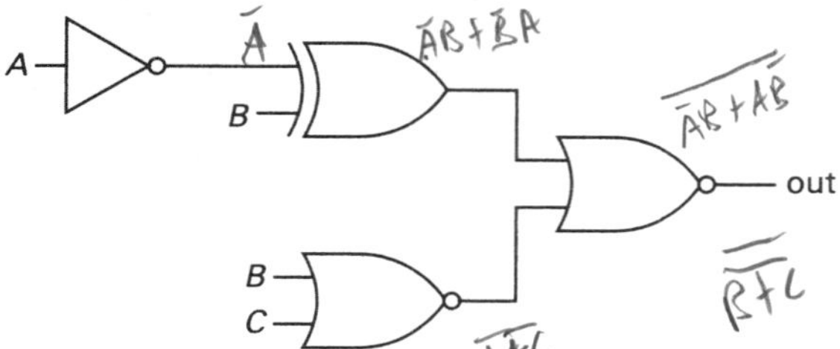

What is the output?

(A) 0

(B) $A\cdot\overline{C}+\overline{B}\cdot C$

(C) $A\cdot\overline{B}\cdot C+\overline{A}\cdot B$

(D) $A\cdot\overline{B}+\overline{A}\cdot B+B\cdot C$

SOLUTIONS

1. The NAND gate at the top left has inputs 1 and 1, so the condition is not satisfied and the output is 0.

The XOR gate at the middle left has inputs 1 and 1, so the condition is not satisfied and the output is 0.

The AND gate at the bottom left has inputs 0 and 0, so the condition is not satisfied and the output is 0.

The XOR gate at the top right has inputs 0 and 0, so the condition is not satisfied and the output is $A=0$.

The NOR gate at the bottom right has inputs 0 and 0, so the condition is satisfied and the output is $B=1$.

The answer is (B).

2. Determine the Boolean logic expression from the logic circuit.

$$\text{out} = A\cdot C\cdot(\overline{0+A\oplus B})$$

The logic "0" on the NOR gate makes the NOR gate effectively an inverter gate.

$$\overline{0+A\oplus B} = \overline{A\oplus B}$$

The output is

$$\text{out} = A\cdot C\cdot(\overline{A\oplus B})$$

The EXCLUSIVE-OR is equivalent to the conditions that will make the output true when the inputs are different.

$$A\oplus B = A\cdot\overline{B}+\overline{A}\cdot B$$

The negation of the EXCLUSIVE-OR is

$$\overline{A\oplus B} = \overline{A\cdot\overline{B}+\overline{A}\cdot B}$$

Apply De Morgan's theorem to the negation of the EXCLUSIVE-OR.

$$\begin{aligned}\overline{A\oplus B} &= \overline{A\cdot\overline{B}+\overline{A}\cdot B} = (\overline{A\cdot\overline{B}})\cdot(\overline{\overline{A}\cdot B})\\ &= (\overline{A}+B)\cdot(A+\overline{B})\end{aligned}$$

The AND is distributive over the OR, so the terms can be distributed analogously to cross multiplication in regular algebra.

$$\begin{aligned}\overline{A\oplus B} &= (\overline{A}+B)\cdot(A+\overline{B})\\ &= A\cdot\overline{A}+A\cdot B+\overline{A}\cdot\overline{B}+B\cdot\overline{B}\\ &= 0+A\cdot B+\overline{A}\cdot\overline{B}+0\\ &= A\cdot B+\overline{A}\cdot\overline{B}\end{aligned}$$

Substitute this result into the equation for the output and simplify.

$$\begin{aligned}\text{out} &= A\cdot C\cdot(\overline{A\oplus B})\\ &= A\cdot C\cdot(A\cdot B+\overline{A}\cdot\overline{B})\\ &= A\cdot C\cdot A\cdot B+A\cdot C\cdot\overline{A}\cdot\overline{B}\\ &= A\cdot A\cdot B\cdot C+A\cdot\overline{A}\cdot\overline{B}\cdot C\\ &= A\cdot B\cdot C+0\cdot\overline{B}\cdot C\\ &= A\cdot B\cdot C\end{aligned}$$

The answer is (A).

3. The Boolean equation for the system can be written from the diagram.

$$\text{out} = (A+\overline{A}\cdot B)\cdot(\overline{D}+C\cdot\overline{C})\cdot\big(C\cdot(C+D)\big)$$

If $A=1$, then $A+\overline{A}\cdot B=1+0\cdot B=1$, regardless of the value of B. If $A=0$, then $A+\overline{A}\cdot B=0+1\cdot B=B$. Therefore, $A+\overline{A}\cdot B=A+B$.

$$\text{out} = (A+B)\cdot(\overline{D}+C\cdot\overline{C})\cdot\big(C\cdot(C+D)\big)$$

$C\cdot\overline{C}=0$, so

$$\begin{aligned}\text{out} &= (A+B)\cdot(\overline{D}+0)\cdot\big(C\cdot(C+D)\big)\\ &= (A+B)\cdot\overline{D}\cdot\big(C\cdot(C+D)\big)\end{aligned}$$

The last product, $C\cdot(C+D)$, is equal to 1 when $C=1$ and is equal to 0 when $C=0$, regardless of the value of D, so

$$\text{out} = (A+B)\cdot\overline{D}\cdot C$$

The output is 1 only when $C=1$, $D=0$, and either $A=1$ or $B=1$; the output is 0 under all other conditions. Only column III shows this pattern.

The answer is (C).

4. From De Morgan's second theorem, the output of the NAND gate is

$$\overline{A\cdot B} = \overline{A}+\overline{B}$$

From De Morgan's first theorem, the output of the NOR gate is

$$\overline{A+B+C} = \overline{A}\cdot\overline{B}\cdot\overline{C}$$

The output of the AND gate combines the output of the other two gates.

$$\text{out} = (\overline{A}+\overline{B})(\overline{A}\cdot\overline{B}\cdot\overline{C})$$

Use the distributive law.

$$\begin{aligned}\text{out} &= \big(\overline{A}\cdot(\overline{A}\cdot\overline{B}\cdot\overline{C})\big)+\big(\overline{B}\cdot(\overline{A}\cdot\overline{B}\cdot\overline{C})\big)\\ &= \big((\overline{A}\cdot\overline{A})\cdot\overline{B}\cdot\overline{C}\big)+\big(\overline{A}\cdot(\overline{B}\cdot\overline{B})\cdot\overline{C}\big)\\ &= (\overline{A}\cdot\overline{B}\cdot\overline{C})+(\overline{A}\cdot\overline{B}\cdot\overline{C})\\ &= \overline{A}\cdot\overline{B}\cdot\overline{C}\end{aligned}$$

The answer is (D).

5. The outputs from the three NOR gates from top to bottom are

$$\overline{A+B+C}$$

$$\overline{A+B+\overline{C}}$$

$$\overline{\overline{A}+B+\overline{D}}$$

The output from the NAND gate is

$$Z = \overline{\overline{(A+B+C)}\cdot\overline{(A+B+\overline{C})}\cdot\overline{(\overline{A}+B+\overline{D})}}$$

Applying De Morgan's second theorem gives

$$\begin{aligned}Z &= \overline{\overline{(A+B+C)}}+\overline{\overline{(A+B+\overline{C})}}+\overline{\overline{(\overline{A}+B+\overline{D})}}\\ &= (A+B+C)+(A+B+\overline{C})+(\overline{A}+B+\overline{D})\\ &= 1\end{aligned}$$

The answer is (B).

6. The Boolean expression for the circuit is

$$\text{out} = \overline{\overline{A}\oplus B+\overline{B+C}}$$

Applying De Morgan's first theorem gives

$$\begin{aligned}\text{out} &= \overline{(\overline{A}\oplus B)}\cdot\overline{\overline{(B+C)}}\\ &= \overline{(\overline{A}\oplus B)}\cdot(B+C)\end{aligned}$$

The result of the EXCLUSIVE-OR is true when the inputs are different.

$$\begin{aligned}\overline{A}\oplus B &= \overline{A}\cdot\overline{B}+\overline{\overline{A}}\cdot B\\ &= \overline{A}\cdot\overline{B}+A\cdot B\end{aligned}$$

The negation of the EXCLUSIVE-OR is

$$\overline{\overline{A}\oplus B} = \overline{\overline{A}\cdot\overline{B}+A\cdot B}$$

Digital Systems

Applying De Morgan's first theorem gives

$$\begin{aligned}\overline{\overline{A} \oplus B} &= \overline{(\overline{A} \cdot \overline{B})} \cdot \overline{(A \cdot B)} \\ &= (\overline{\overline{A}} + \overline{\overline{B}}) \cdot (\overline{A} + \overline{B}) \\ &= (A + B) \cdot (\overline{A} + \overline{B})\end{aligned}$$

The distributive law gives

$$\begin{aligned}\overline{\overline{A} \oplus B} &= A \cdot \overline{A} + A \cdot \overline{B} + B \cdot \overline{A} + B \cdot \overline{B} \\ &= 0 + A \cdot \overline{B} + B \cdot \overline{A} + 0 \\ &= A \cdot \overline{B} + \overline{A} \cdot B\end{aligned}$$

Substitute this result into the equation for the output and use the distributive law.

$$\begin{aligned}\text{out} &= (A \cdot \overline{B} + \overline{A} \cdot B) \cdot (B + C) \\ &= A \cdot \overline{B} \cdot B + A \cdot \overline{B} \cdot C + \overline{A} \cdot B \cdot B + \overline{A} \cdot B \cdot C \\ &= 0 + A \cdot \overline{B} \cdot C + \overline{A} \cdot B + \overline{A} \cdot B \cdot C \\ &= A \cdot \overline{B} \cdot C + \overline{A} \cdot B + \overline{A} \cdot B \cdot C\end{aligned}$$

Use the distributive law to factor $\overline{A} \cdot B$ out of the second and third terms.

$$\begin{aligned}\text{out} &= A \cdot \overline{B} \cdot C + (\overline{A} \cdot B) \cdot (1 + C) \\ &= A \cdot \overline{B} \cdot C + \overline{A} \cdot B \cdot 1 \\ &= A \cdot \overline{B} \cdot C + \overline{A} \cdot B\end{aligned}$$

The answer is (C).

38 Logic Network Design

PRACTICE PROBLEMS

1. The following Boolean function is in sum-of-products (SOP) form.

$$Y = A \cdot \overline{B} \cdot C + A \cdot \overline{B} \cdot \overline{C} + \overline{A} \cdot \overline{B} \cdot \overline{C} + \overline{A} \cdot \overline{B} \cdot C$$

Which of these represents the same function in product-of-sums (POS) form?

(A) $Y = (A + \overline{B} + C) \cdot (\overline{A} + \overline{B}) \cdot (\overline{B} + \overline{C})$

(B) $Y = (\overline{A} + B + \overline{C}) \cdot (A + B)$

(C) $Y = (\overline{A} + B + \overline{C}) \cdot (A + B) \cdot (\overline{A} + \overline{B} + C)$

(D) $Y = (\overline{A} + B + \overline{C}) \cdot (A + \overline{B} + C) \cdot (B + C)$

2. What terms are essential prime implicants in the given Karnaugh map?

CD \ AB	00	01	11	10
00	0	0	0	0
01	0	X	0	0
11	0	0	0	0
10	0	X	1	1

(A) $A \cdot C \cdot \overline{D}$ only

(B) $B \cdot C \cdot \overline{D}$ only

(C) $A \cdot C \cdot \overline{D}$ and $B \cdot C \cdot \overline{D}$ only

(D) $\overline{A} \cdot B \cdot \overline{C}$, $A \cdot C \cdot \overline{D}$, and $B \cdot C \cdot \overline{D}$ only

3. This Boolean expression for F cannot be further simplified.

$$F = \overline{A} \cdot \overline{B} \cdot \overline{C} + A \cdot \overline{B} \cdot C + \overline{A} \cdot B \cdot C \cdot D$$

Which terms in the expression are the essential prime implicants of F?

(A) $A \cdot \overline{B} \cdot C$ only

(B) $\overline{A} \cdot B \cdot C \cdot D$ only

(C) $\overline{A} \cdot \overline{B} \cdot \overline{C}$ and $A \cdot \overline{B} \cdot C$ only

(D) $\overline{A} \cdot \overline{B} \cdot \overline{C}$, $A \cdot \overline{B} \cdot C$, and $\overline{A} \cdot B \cdot C \cdot D$

4. F is a Boolean function.

$$F(A, B, C) = (A \cdot B \cdot C) + (A + C) + A \cdot B + B$$

Which of the following is a maxterm of F?

(A) $A + \overline{B}$

(B) $A \cdot C$

(C) $A + B + C$

(D) $A \cdot \overline{B} \cdot C$

5. F is a Boolean function.

$$F = (A \cdot B \cdot C) + (A + C) + A \cdot B + B$$

Which of the following is a minterm of F?

(A) $A + \overline{B}$

(B) $A \cdot C$

(C) $A + B + C$

(D) $A \cdot \overline{B} \cdot C$

6. How many prime implicants does the following logic expression have?

$$f(A, B, C, D) = m_4 + m_5 + m_6 + m_7 + m_8 + m_{13} + m_{15}$$

(A) 3

(B) 4

(C) 7

(D) 16

7. Find the logic expression which implements the Boolean function $F(A, B, C) = \bar{A}\cdot\bar{B}\cdot\bar{C} + B\cdot\bar{C} + A\cdot B\cdot C$ using a minimum number of NAND gates only, in a hazard-free circuit.

(A) $(\bar{A}\cdot\bar{C})\cdot(A\cdot B)$

(B) $\overline{\overline{(\bar{A}\cdot\bar{C})}\cdot\overline{(B\cdot\bar{C})}\cdot\overline{(A\cdot B)}}$

(C) $\overline{\overline{(\bar{A}\cdot\bar{C})}\cdot\overline{(A\cdot B)}}$

(D) $(\bar{A}\cdot\bar{C}) + (B\cdot\bar{C}) + (A\cdot B)$

SOLUTIONS

1. The Boolean function in SOP form contains the minterms 101, 100, 000, and 001. The POS form must contain the maxterms 010, 011, 111, and 110. (There is a total of 2^n terms, where n is the number of variables.)

$$Y = (A + \bar{B} + C)\cdot(A + \bar{B} + \bar{C})\cdot(\bar{A} + \bar{B} + \bar{C})\cdot(\bar{A} + \bar{B} + C)$$

Both conditions of A are present in the maxterms 011 and 111, so they can be combined in the term

$$(A + \bar{B} + \bar{C})\cdot(\bar{A} + \bar{B} + \bar{C}) = \bar{B} + \bar{C}$$

Similarly,

$$(\bar{A} + \bar{B} + C)\cdot(\bar{A} + \bar{B} + \bar{C}) = \bar{A} + \bar{B}$$

The combined result is

$$Y = (A + \bar{B} + C)\cdot(\bar{A} + \bar{B})\cdot(\bar{B} + \bar{C})$$

The answer is (A).

2. A prime implicant is one that is not contained entirely within another implicant. There are three possible prime implicants in this Karnaugh map.

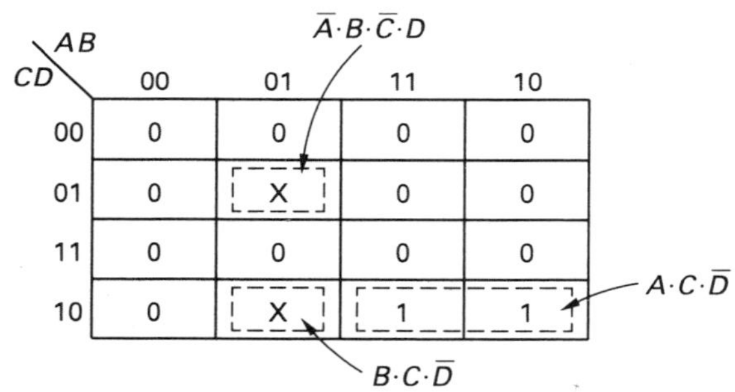

An essential prime implicant is one that is not contained entirely within another prime implicant. Of these three possible prime implicants, two contain "don't cares." These cells cannot be grouped with the 1-cells to make a larger grouping, so in the simplest possible expression they have a value of zero. The remaining prime implicant, $A\cdot C\cdot\bar{D}$ only, covers all the 1-cells, so this is the only essential prime implicant.

The answer is (A).

3. All the terms are essential because the Boolean equation cannot be farther simplified. All essential primes must be part of the minimized expression. This can be visualized by mapping the implicants into a Karnaugh map.

CD \ AB	00	01	11	10
00	1	0	0	0
01	1	0	0	0
11	0	1	0	1
10	0	0	1	1

$\overline{A}\cdot\overline{B}\cdot\overline{C}$

$A\cdot\overline{B}\cdot C$

$\overline{A}\cdot B\cdot C\cdot D$

The answer is (D).

4. By definition, a maxterm is a sum (OR) of all the variables in a function, with each variable used exactly once either directly or in negated form. Among the given options, $A + B + C$ is the only maxterm.

The answer is (C).

5. By definition, a minterm is a product (AND) of all the variables in a function, with each variable used exactly once either directly or in negated form. Among the given options, $A \cdot \overline{B} \cdot C$ is the only minterm.

The answer is (D).

6. Plot the expression on a Karnaugh map to aid in seeing the prime implicants. A prime implicant is one that is not contained entirely within another implicant.

CD \ AB	00	01	11	10
00	M_0	M_1	M_3	M_2
01	M_4	M_5	M_7	M_6
11	M_{12}	M_{13}	M_{15}	M_{14}
10	M_8	M_9	M_{11}	M_{10}

$A \cdot B$, $B \cdot D$, and $A \cdot B \cdot C \cdot D$ are the prime implicants.

The answer is (A).

7. Draw the Karnaugh map to find the sum-of-products form with the smallest number of groups.

CD \ AB	00	01	11	10
00	1	1	1	
01			1	

$$F(A, B, C) = \overline{A} \cdot \overline{C} + A \cdot B$$

Use De Morgan's first theorem to express the result in terms of NAND gates.

$$\begin{aligned}\overline{A} \cdot \overline{C} + A \cdot B &= \overline{\overline{\overline{A} \cdot \overline{C} + A \cdot B}} \\ &= \overline{(\overline{A} \cdot \overline{C}) \cdot \overline{(A \cdot B)}}\end{aligned}$$

The answer is (C).

39 Sequential Networks

PRACTICE PROBLEMS

1. Compared to an SR flip-flop, which additional functions does a JK flip-flop have?

(A) holding the current state

(B) asynchronous set and reset

(C) clock frequency division

(D) inverting the current state

2. Which of the following statements about the use of SR flip-flops are true?

I. They can store an input at a clock edge.

II. They can set one bit of memory to 1.

III. They can reset one bit of memory to 0.

(A) I and II only

(B) I and III only

(C) II and III only

(D) I, II, and III

3. The sequential network circuit in the following diagram starts with all bits equal to 0. The flip-flops are leading edge triggered (with a state transition initiated by a change from 0 to 1 on the clock). The clock period is long compared to the flip-flop transition time. Bit 0 is the least significant bit, and bit 2 is the most significant bit. After three clock pulses, what is the count on the circuit (in the sequence bit 0, bit 1, bit 2)?

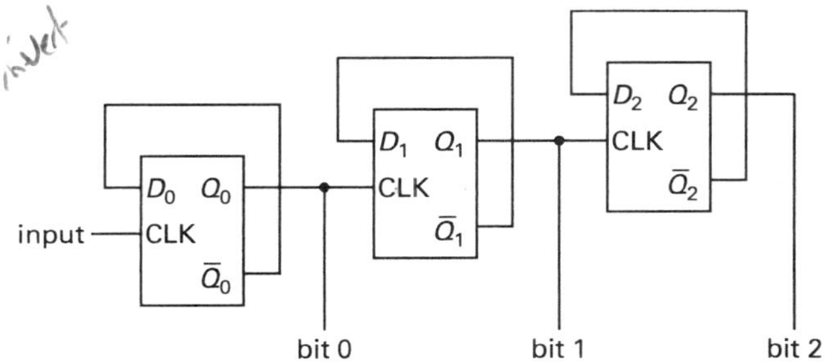

(A) 011

(B) 100

(C) 101

(D) 110

4. The sequential network circuit in the following diagram starts with all bits equal to 0. Bit 0 is the least significant bit and bit 1 is the most significant bit. After four clock pulses, what is the count on the circuit (in the sequence bit 0, bit 1)?

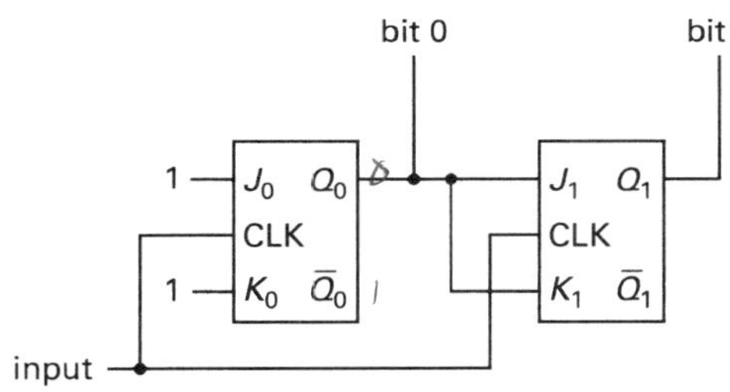

(A) 00

(B) 01

(C) 10

(D) 11

5. The sequential network circuit shown starts with flip-flop 0 equal to 1 and flip-flop 1 equal to 1. Bit 0 is the least significant bit and bit 1 is the most significant bit. After one clock pulse, what is the count on the circuit (in the sequence bit 0, bit 1)?

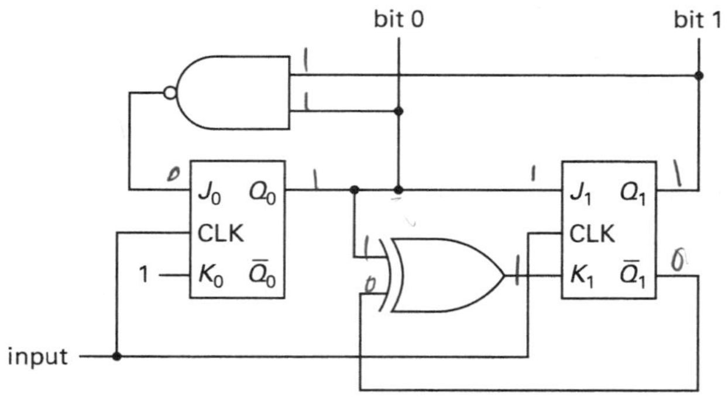

(A) 00

(B) 01

(C) 10

(D) 11

6. The sequential network circuit in the following diagram starts with flip-flop 0 equal to 1, flip-flop 1 equal to 1, and flip-flop 2 equal to 0. Bit 0 is the least significant bit and bit 2 is the most significant bit. After one clock pulse, what is the output count on the circuit (in the sequence bit 0, bit 1, bit 2)?

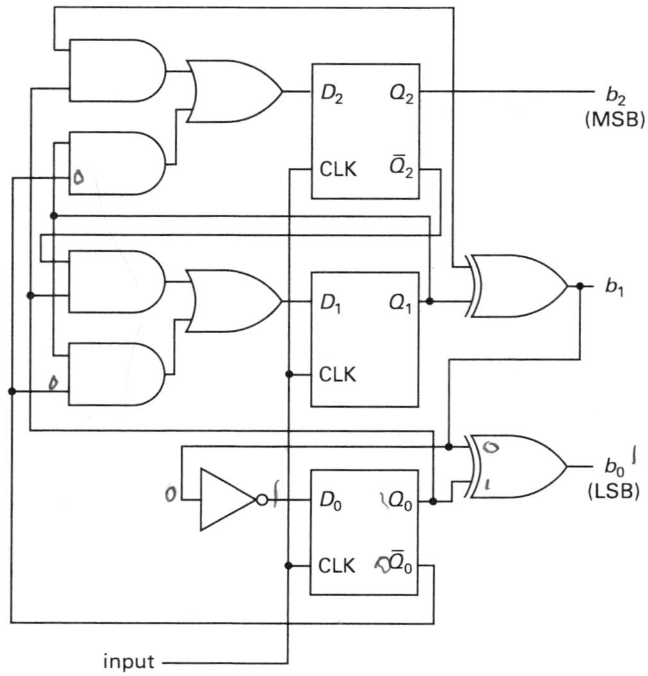

Adapted from *Computer Engineering Reference Manual for the Electrical and Computer PE Exam*, by John A. Camara, PE, copyright © 2010, by Professional Publications, Inc.

(A) 001

(B) 100

(C) 110

(D) 111

SOLUTIONS

1. A JK flip-flop performs the same functions as a SR flip-flop. In addition, a JK flip-flop can toggle or invert the current state when both inputs are true. An SR flip-flop cannot do this because having both inputs true is an invalid condition for an SR flip-flop.

Both SR and JK flip-flops will hold the current state if both inputs are false. Option A is incorrect.

Both SR and JK flip-flops can have asynchronous inputs for set and reset of the flip-flop. Option B is incorrect.

Both SR and JK flip-flops can be used in frequency division networks. Option C is incorrect.

The answer is (D).

2. An SR flip-flop sets or resets the flip-flop state based on the inputs.

- If S is 1 and R is 0, the flip-flop state is 1 following the appropriate clock edge.
- If R is 1 and S is 0, the flip-flop state is 0 following the appropriate clock edge.

The answer is (D).

3. The first clock pulse will cause flip-flop 0 to transition. The next state of flip-flop 0 is

$$Q_0^+ = D_0 = \overline{Q}_0 = 1$$

This results in a transition of flip-flop 1, because its clock input changes from 0 to 1. The output of flip-flop 1 becomes

$$Q_1^+ = D_1 = \overline{Q}_1 = 1$$

This results in a transition of flip-flop 2, because its clock input changes from 0 to 1. The output of flip-flop 2 becomes

$$Q_2^+ = D_2 = \overline{Q}_2 = 1$$

The second clock pulse will cause flip-flop 0 to transition. The next state of flip-flop 0 is

$$Q_0^+ = D_0 = \overline{Q}_0 = 0$$

This transition from 1 to 0 does not cause flip-flop 1 to transition, so flip-flop 2 also does not transition.

The third clock pulse will cause flip-flop 0 to transition. The next state of flip-flop 0 is

$$Q_0^+ = D_0 = \overline{Q}_0 = 1$$

This results in a transition of flip-flop 1, because its clock input changes from 0 to 1. The output of flip-flop 1 becomes

$$Q_1^+ = D_1 = \overline{Q}_1 = 0$$

This transition from 1 to 0 does not cause flip-flop 2 to transition.

The count after three clock pulses is $Q_0 = 1$, $Q_1 = 0$, and $Q_2 = 1$, or 101.

The answer is (C).

4. The inputs to flip-flop 0 are both true, so that flip-flop will toggle each clock pulse. The first clock pulse will cause flip-flop 0 to transition. The next state of flip-flop 0 is

$$Q_0^+ = \overline{Q}_0 = 1$$

The inputs of flip-flop 1 are both false at the first clock pulse, so flip-flop 1 holds the current condition.

$$Q_1^+ = \overline{Q}_1 = 0$$

Flip-flop 0 will toggle again after the second clock pulse. T, the next state of flip-flop 0 is

$$Q_0^+ = \overline{Q}_0 = 0$$

The inputs of flip-flop 1 are both true at the second clock pulse, so flip-flop 1 toggles.

$$Q_1^+ = \overline{Q}_1 = 1$$

Flip-flop 0 will toggle again after the third clock pulse. The next state of flip-flop 0 is

$$Q_0^+ = \overline{Q}_0 = 1$$

The inputs of flip-flop 1 are both false at the third clock pulse, so flip-flop 1 holds the current condition.

$$Q_1^+ = \overline{Q}_1 = 1$$

Flip-flop 0 will toggle again after the fourth clock pulse. The next state of flip-flop 0 is

$$Q_0^+ = \overline{Q}_0 = 0$$

The inputs of flip-flop 1 are both true at the fourth clock pulse, so flip-flop 1 toggles.

$$Q_1^+ = \overline{Q}_1 = 0$$

The count after four clock pulses is $Q_1 = 0$ and $Q_0 = 0$, or 00.

The answer is (A).

5. The diagram shows the current state conditions of the circuit before the clock pulse.

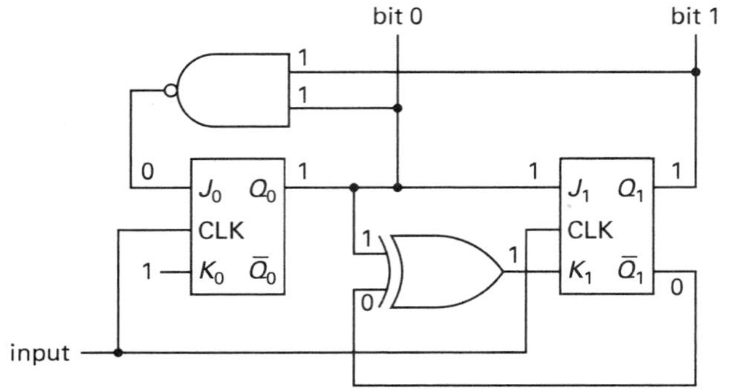

For flip-flop 0:

$$J_0 = 0$$
$$K_0 = 1$$
$$Q(t+1) = 0$$

For flip-flop 1:

$$S_1 = 1$$
$$R_1 = 1$$
$$Q(t+1) = Q(t) = 1$$

The answer is (B).

6. The diagram shows the current state conditions of the circuit before the clock pulse.

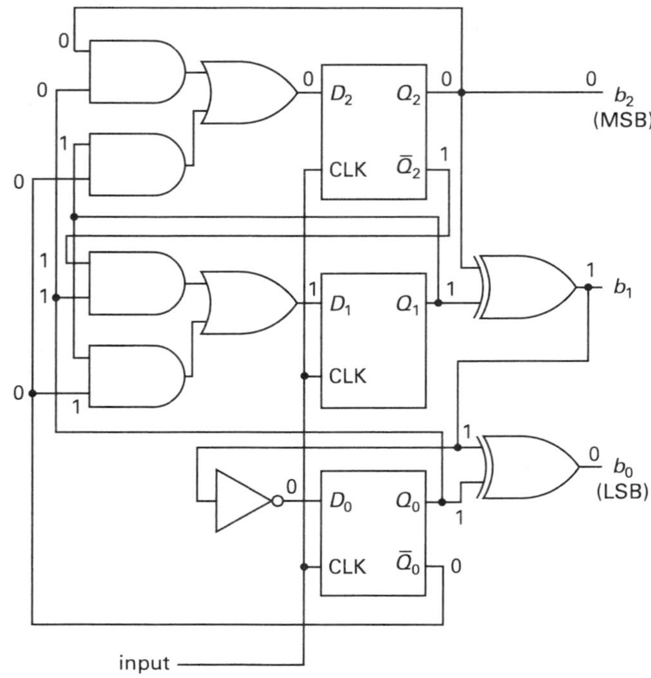

Adapted from *Computer Engineering Reference Manual for the Electrical and Computer PE Exam,* by John A. Camara, PE, copyright © 2010, by Professional Publications, Inc.

The next state of the D flip-flop is the input in the current state condition.

$$D_0^+ = D_0 = 0$$
$$D_1^+ = D_1 = 1$$
$$D_2^+ = D_2 = 0$$

The outputs are

$$b_0 = Q_2 \oplus Q_1 \oplus Q_0 = 0 \oplus 1 \oplus 0 = 1$$
$$b_1 = Q_2 \oplus Q_1 = 0 \oplus 1 = 1$$
$$b_2 = Q_2 = 0$$

The answer is (C).

40 Digital Systems

PRACTICE PROBLEMS

1. A 24 kHz, 16-bit, analog-to-digital converter samples an analog signal. The voltage resolution is 1 mV, and the low end of the voltage range is 0 V. Most nearly, the high end of the voltage range that can be measured is

(A) 18 V

(B) 44 V

(C) 66 V

(D) 120 V

2. A 24 kHz, 16-bit, analog-to-digital converter samples an analog signal. The voltage resolution is 1 mV, and the low end of the voltage range is 0 V. Most nearly, an output of 1100100 from the converter represents a voltage of

(A) 1 mV

(B) 10 mV

(C) 100 mV

(D) 1000 mV

3. An analog signal varying from 0 V to 10 V is sampled by an 8-bit analog-to-digital converter. Most nearly, the voltage resolution is

(A) 0.039 V

(B) 0.33 V

(C) 1.0 V

(D) 1.2 V

4. An 8-bit programmable module will be used to control temperature. If the temperature sensor's range is 0–300°F and the controller's range is 0–5 V_{DC}, what is the resolution capacity of the controller?

(A) 0.02 °F/step

(B) 1.2 °F/step

(C) 2.3 °F/step

(D) 60 °F/step

5. An analog-to-digital converter has a range of −5 V to 5 V and an 8-bit output. Most nearly, an output of 01101100 from the converter corresponds to an input voltage of

(A) −0.82 V

(B) −0.78 V

(C) −0.64 V

(D) 0.039 V

6. A resistance temperature detector (RTD) transducer is to be chosen such that the analog-to-digital conversion will have a resolution of 0.001°C. The conversion uses 16 bits. The RTD will be in a circuit with a lower voltage of −0.150 V at the lowest temperature extreme and an upper voltage of 0.300 V at the highest temperature extreme. The RTD will have an output of 0.000 V at 0.000°C. What is most nearly the sensitivity of the RTD, and what is most nearly the temperature range that the analog-to-digital conversion can represent?

(A) 96.0 °C/V, −43.9°C to 98.6°C

(B) 146 °C/V, −21.9°C to 43.7°C

(C) 160 °C/V, −11.7°C to 38.8°C

(D) 222 °C/V, 1.90°C to 78.7°C

SOLUTIONS

1. Rearrange the voltage resolution equation to find the high voltage.

$$\begin{aligned}\varepsilon_V &= \frac{V_H - V_L}{2^n} \\ V_H &= 2^n\varepsilon_V + V_L = (2)^{16}(1\times 10^{-3}\ \text{V}) + 0\ \text{V} \\ &= 65.54\ \text{V} \quad (66\ \text{V})\end{aligned}$$

The answer is (C).

2. Convert the binary value to decimal form.

$$\begin{aligned}(1100100)_2 &= (1)(2)^6 + (1)(2)^5 + (0)(2)^4 + (0)(2)^3 \\ &\quad + (1)(2)^2 + (0)(2)^1 + (0)(2)^0 \\ &= (100)_{10}\end{aligned}$$

Use the equation for converted voltage to find the voltage represented by this decimal value.

$$\begin{aligned}V &= \varepsilon_V N + V_L = (1\ \text{mV})(100) + 0\ \text{V} \\ &= 100\ \text{mV}\end{aligned}$$

The answer is (C).

3. The voltage resolution is

$$\varepsilon_V = \frac{V_H - V_L}{2^n} = \frac{10\ \text{V} - 0\ \text{V}}{(2)^8} = 0.039\ \text{V}$$

The answer is (A).

4. The number of distinct states, or steps, that the controller can be in is $(2)^8$ or 256. The monitored temperature can range from 0°F to 300°F, so the resolution capacity is

$$\frac{300°\text{F} - 0°\text{F}}{256\ \text{steps}} = 1.17\ °\text{F/step} \quad (1.2\ °\text{F/step})$$

The answer is (B).

5. The number of distinct steps (levels) of the 8-bit digital system is $(2)^8$, or 256. The voltage range is 5 V – (–5 V) = 10 V, so the resolution is

$$\varepsilon_V = \frac{V_H - V_L}{2^n} = \frac{5\ \text{V} - (-5\ \text{V})}{(2)^8\ \text{steps}} = 0.0391\ \text{V/step}$$

The decimal equivalent of $(01101100)_2$ gives the number of steps from the lower voltage range.

$$\begin{aligned}(01101100)_2 &= (0)(2)^7 + (1)(2)^6 + (1)(2)^5 + (0)(2)^4 \\ &\quad + (1)(2)^3 + (1)(2)^2 + (0)(2)^1 + (0)(2)^0 \\ &= (108)_{10}\end{aligned}$$

There are 108 steps. The voltage deviation from the lower voltage range is

$$\left(0.0391\ \frac{\text{V}}{\text{step}}\right)(108\ \text{steps}) = 4.2228\ \text{V}$$

The input voltage is

$$-5\ \text{V} + 4.2228\ \text{V} = -0.7772\ \text{V} \quad (-0.78\ \text{V})$$

The answer is (B).

6. The resolution of the circuit is

$$\begin{aligned}\varepsilon_V &= \frac{V_H - V_L}{2^n} = \frac{0.300\ \text{V} - (-0.150\ \text{V})}{2^{16}} \\ &= 6.8665\times 10^{-6}\ \text{V}\end{aligned}$$

This voltage is to represent 0.001°C with the RTD; therefore, the sensitivity of the RTD should be

$$\frac{0.001°\text{C}}{6.8665\times 10^{-6}\ \text{V}} = 145.64\ °\text{C/V} \quad (146\ °\text{C/V})$$

The temperature range that this sensor can represent is derived from the number of bits and the value that represents 0°C. The voltage range is

$$(0.001°\text{C})(2^{16}) = 65.54°\text{C}$$

The value that represents 0°C is 0 V, and one-third of the voltage range is below zero while two-thirds of the voltage range is above zero. The lower and upper limits the sensor can represent are

$$\frac{-65.54°\text{C}}{3} = -21.85°\text{C} \quad (-21.9°\text{C})$$

$$\frac{(2)(65.54°\text{C})}{3} = 43.69°\text{C} \quad (43.7°\text{C})$$

The answer is (B).

41 Computer Software

PRACTICE PROBLEMS

1. Which of the following flowcharts does NOT represent a complete program?

(A)

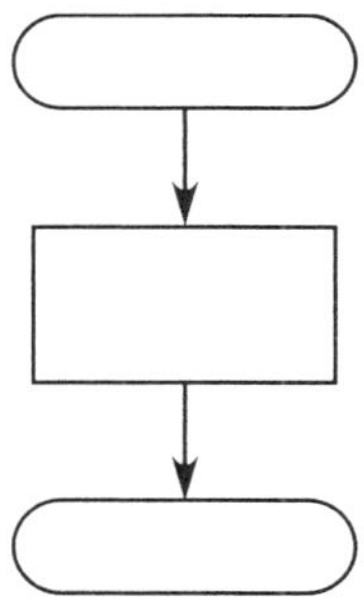

(B)

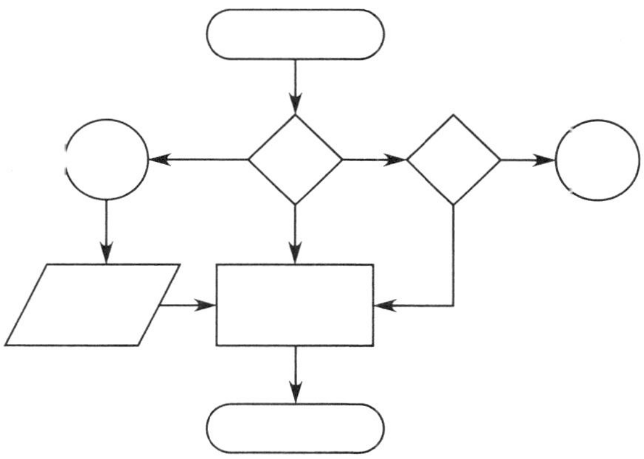

(C)

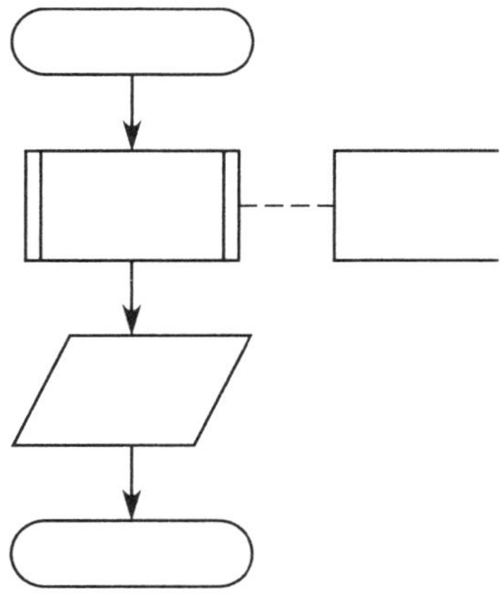

(D)

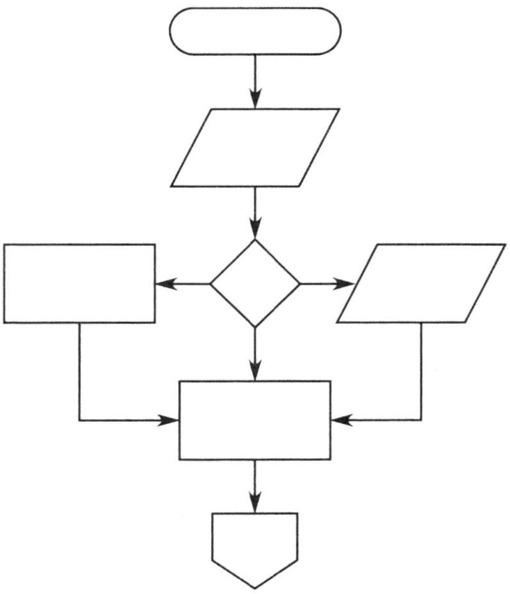

Software Development

2. What flowchart element is used to represent an IF... THEN statement?

(A)

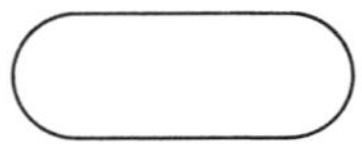

(B)

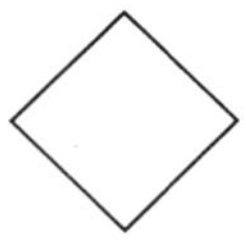

(C)

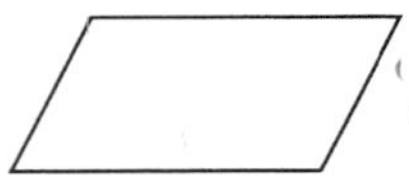

(D)

3. In programming, a recursive function is a function that

(A) calls previously used functions

(B) generates functional code to replace symbolic code

(C) calls itself

(D) compiles itself in real time

4. Structured programming is to be used to determine whether examinees pass a test. A passing score is 70 or more out of a possible 100. Which of the following IF statements would set the variable PASSED to 1 (true) when the variable SCORE is passing, and set the variable PASSED to 0 (false) when the variable SCORE is not passing?

(A) IF SCORE > 70 PASSED = 1
ELSE PASSED = 0

(B) IF SCORE > 69 PASSED = 1
ELSE PASSED = 0

(C) IF SCORE < 69 PASSED = 1

(D) IF SCORE < 69 PASSED = 0
ELSE PASSED = 1

5. A structured programming segment is shown.

```
Y = 4
B = 4
Y = 3 * B - 6
IF Y > B THEN Y = B - 2
IF Y < B THEN Y = Y + 2
IF Y = B THEN Y = B + 2
```

What is the value of Y after the segment is executed?

(A) 2

(B) 6

(C) 8

(D) 12

6. A structured programming segment is shown. The variable N is an integer greater than zero.

```
A = X
DO UNTIL N = 0
   Y = A * X
   A = Y
   N = N - 1
END UNTIL
```

Which equation is implemented by this programming segment?

(A) $Y = X!$

(B) $Y = X^{N-1}$

(C) $Y = X^{N}$

(D) $Y = X^{N+1}$

7. A structured programming fragment is shown.

```
1 REAL X, Y
2 X = 3
3 Y = COS(X)
4 PRINT Y
```

Line 2 is

(A) an assignment

(B) a command

(C) a declaration

(D) a function

Software Development

8. In a typical spreadsheet program, what cell is directly below cell AB4?

(A) AB5

(B) AC4

(C) AC5

(D) BC4

9. Which of the following terms is best defined as a formula or set of steps for solving a particular problem?

(A) program

(B) software

(C) firmware

(D) algorithm

10. The computer language that is executed within a computer's central processing unit is called

(A) MS-DOS

(B) a high-level language

(C) an assembly language

(D) a machine language

11. Which of the following best defines a compiler?

(A) hardware that is used to translate high-level language to machine code

(B) software that collects and stores executable commands in a program

(C) software that is used to translate high-level language into machine code

(D) hardware that collects and stores executable commands in a program

12. The effect of using recursive functions in a program is generally to use

(A) less code and less memory

(B) less code and more memory

(C) more code and less memory

(D) more code and more memory

13. In which of these situations can an 8-bit system correctly access more than 128 different integers?

(A) when the integers are in the range of [−255, 0]

(B) when the integers are in the range of [0, 256]

(C) when the integers are in the range of [−128, 128]

(D) when the integers are in the range of [0, 512]

14. In which of these computer operating systems can a document in HTML (hypertext markup language) format be viewed?

I. Mac OS

II. Linux

III. Windows

IV. Unix

(A) I and III only

(B) I, II, and III only

(C) I, III, and IV only

(D) I, II, III, and IV

15. A typical spreadsheet for economic evaluation of alternatives uses cell F4 to store the percentage value of inflation rate. The percentage rate is assumed to be constant throughout the lifetime of the study. What variable should be used to access that value throughout the model?

(A) F4

(B) \$F4

(C) %F4

(D) \$F\$4

16. Refer to the following portion of a spreadsheet.

	A	B	C
1	10	11	12
2	1	A2^2	
3	2	A3^2	
4	3	A4^2	
5	4	A5^2	

The top-to-bottom values in column B will be

(A) 11, 1, 2, 3, 4

(B) 11, 1, 3, 6, 10

(C) 11, 1, 4, 9, 16

(D) 11, 1, 5, 12, 22

17. Refer to the following portion of a spreadsheet.

	A	B	C	D
1	10	11	12	13
2	5	4	B2*A\$1	
3	6	5	B3*B\$1	
4	7	6	B4*C\$1	
5	8	7	B5*D\$1	

The top-to-bottom values in column C will be

(A) 12, 20, 30, 42, 56

(B) 12, 40, 55, 72, 91

(C) 12, 50, 66, 84, 104

(D) 12, 100, 121, 144, 169

18. What is the value of J after the given program segment is executed?

```
J = 0
WHILE (J < 10) DO
J = J + 3
END
```

(A) 3

(B) 6

(C) 9

(D) 12

SOLUTIONS

1. A flowchart must begin and end with a terminal symbol. The symbol at the bottom of option D is the "off-page" symbol, which indicates that the flowchart continues on the next page. This is not a complete program.

The answer is (D).

2. At an IF...THEN statement, the flow of a program is decided based on a criterion that can be evaluated as true or false. The symbol used to represent a decision is the diamond.

The answer is (B).

3. A recursive function calls itself.

The answer is (C).

4. Option B sets PASSED to 1 for SCORE = 70 to 100 and sets PASSED to 0 for SCORE = 0 to 69.

Option A will not give the correct response when SCORE = 70. Option C will never set PASSED to 0. Option D will not give the correct response when SCORE = 69.

The answer is (B).

5. The first operation changes the value of Y.

$$Y = 3 \times 4 - 6 = 6$$

The first IF statement is satisfied, so the operation is performed.

$$Y = 4 - 2 = 2$$

However, the program execution does not end here.

The value of Y is then less than B, so the second IF statement is executed. This statement is satisfied, so the operation is performed.

$$Y = 2 + 2 = 4$$

The value of Y is then equal to B, so the third IF statement is executed. This statement is satisfied, so the operation is performed.

$$Y = 4 + 2 = 6$$

The answer is (B).

6. The DO/UNTIL loop will be executed N times. After one execution, $Y = X^2$. Each subsequent execution of the loop multiplies Y by X another time. Therefore, this segment calculates $Y = X^{N+1}$.

The answer is (D).

7. Line 2 is an assignment. A command, such as line 4, directs the computer to take some action such as PRINT. A declaration, such as line 1, states what type of data a variable will contain (like REAL) and reserves space for it in memory. A function, such as line 3, performs a specific operation (such as finding the cosine of a number) and returns a value.

The answer is (A).

8. Spreadsheets generally label a cell by giving its column and row, in that order. Cell AB4 is in column AB, row 4. The cell directly below AB4 is in column AB, row 5, designated as AB5.

The answer is (A).

9. An algorithm is a formula or set of steps for solving a particular problem.

A program is a sequence of instructions that implements a formula or set of steps, but the program is not itself the formula or set of steps. An algorithm is often implemented as a program. Software and firmware are programs stored on media.

The answer is (D).

10. The central processing unit executes a version of the program that has been compiled into the machine language. This version of the program is in the form of operations and operands specific to the machine's coding.

The answer is (D).

11. A compiler is a program (i.e., software) that converts programs written in higher-level languages to lower-level languages that the computer can understand.

The answer is (C).

12. A recursive function calls itself. Since the function does not need to be coded in multiple places, less code is used. Each subsequent call of the function must be carried out in a different location, so more memory is used.

The answer is (B).

13. An 8-bit system can represent $(2)^8 = 256$ different distinct integers. Normally, the eighth bit is used for the sign, and only seven bits are used for magnitude, resulting in a range of $[-127, 128]$ or $[-128, 127]$ (counting zero as one of the integers). If all the integers are known or assumed to have the same sign, a range of 256 integers is available. All the answer options except option A contain more than 256 distinct integers.

The answer is (A).

14. HTML may be viewed on any computer with a compatible browser.

The answer is (D).

15. The dollar sign symbol, "$", is used in spreadsheets to "fix" the column and/or row designator following it when other columns or rows are permitted to vary.

The answer is (D).

16. Except for the first entry (which is 11), column B calculates the square of the values in column A. The entries are 11, $(1)^2$, $(2)^2$, $(3)^2$, and $(4)^2$.

The answer is (C).

17. Except for the first entry (which is 12), column C is found by taking the numbers from column B and then multiplying by the entries in row 1. For example, B2*A$1 means to multiply the entry in B2, which is 4, by the number entered in cell A1, which is 10. This product is 40.

The entries are 12, 4×10, 5×11, 6×12, 7×13, or 12, 40, 55, 72, 91.

The answer is (B).

18. Since the control will exit the DO loop when J becomes larger than or equal to 10, J has the following values: 0, 3, 6, 9, and 12. When J is 12, the program ends.

The answer is (D).

42 Engineering Economics

PRACTICE PROBLEMS

1. A computer with a useful life of five years has an initial cost of \$6000. The salvage value is \$2300, and the annual maintenance is \$210/yr. The interest rate is 8%. What is most nearly the present worth of the costs for the computer?

(A) \$5200

(B) \$5300

(C) \$5600

(D) \$5700

2. A company purchases a piece of equipment for \$15,000. After nine years, the salvage value is \$900. The annual insurance cost is 5% of the purchase price, the electricity cost is \$600/yr, and the maintenance and replacement parts cost is \$120/yr. The effective annual interest rate is 10%. Neglecting taxes, what is most nearly the present worth of the equipment if it is expected to save the company \$4500 per year?

(A) \$2300

(B) \$2800

(C) \$3200

(D) \$3500

3. A company must purchase a machine that will be used over the next eight years. The purchase price is \$10,000, and the salvage value after eight years is \$1000. The annual insurance cost is 2% of the purchase price, the electricity cost is \$300 per year, and maintenance and replacement parts cost \$100 per year. The effective annual interest rate is 6%. Neglect taxes. Most nearly, what is the effective uniform annual cost (EUAC) of ownership?

(A) \$1200

(B) \$2100

(C) \$2200

(D) \$2300

4. An oil company is planning to install a new 80 mm pipeline to connect storage tanks to a processing plant 1500 m away. The connection will be needed for the foreseeable future. An annual interest rate of 8% is assumed, and annual maintenance and pumping costs are considered to be paid in their entireties at the end of the years in which their costs are incurred.

initial cost	\$1500
service life	12 yr
salvage value	\$200
annual maintenance	\$400
pump cost/hour	\$2.50
pump operation	600 hr/yr

Most nearly, what is the capitalized cost of running and maintaining the 80 mm pipeline?

(A) \$15,000

(B) \$20,000

(C) \$24,000

(D) \$27,000

5. New 200 mm diameter pipeline is installed over a distance of 1000 m. Annual maintenance and pumping costs are considered to be paid in their entireties at the end of the years in which their costs are incurred. The pipe has the following costs and properties.

initial cost	\$1350
annual interest rate	6%
service life	6 yr
salvage value	\$120
annual maintenance	\$500
pump cost/hour	\$2.75
pump operation	2000 hr/yr

What is most nearly the equivalent uniform annual cost (EUAC) of the pipe?

(A) \$5700

(B) \$5900

(C) \$6100

(D) \$6300

6. A piece of equipment has an initial cost of \$5000 in year 1. The maintenance cost is \$300/yr for the total lifetime of seven years. During years 1–3, the rate of inflation is 5%, and the effective annual rate of interest is 9%. Most nearly, what is the uninflated present worth of the equipment during year 1?

(A) \$3200

(B) \$3300

(C) \$3400

(D) \$3500

7. A computer with a useful life of 13 years has the following costs and interest rate.

initial cost	\$5500
salvage value	\$3100
annual maintenance	
years 1–8	\$275
years 9–13	\$425
interest rate	6%

Most nearly, what is the equivalent uniform annual cost (EUAC) of the computer?

(A) \$730

(B) \$780

(C) \$820

(D) \$870

8. Permanent mineral rights on a parcel of land are purchased for an initial lump-sum payment of \$100,000. Profits from mining activities are \$12,000 each year, and these profits are expected to continue indefinitely. Most nearly, what is the interest rate earned on the initial investment?

(A) 8.3%

(B) 9.0%

(C) 10%

(D) 12%

9. \$1000 is deposited in a savings account that pays 6% annual interest, and no money is withdrawn for three years. Most nearly, what is the account balance after three years?

(A) \$1120

(B) \$1190

(C) \$1210

(D) \$1280

10. A construction company purchases 100 m of 40 mm diameter steel cable with an initial cost of \$4500. The annual interest rate is 4%, and annual maintenance costs are considered to be paid in their entireties at the end of the years in which their costs are incurred. The annual maintenance cost of the cable is \$200/yr over a service life of nine years. Using Modified Accelerated Cost Recovery System (MACRS) depreciation and assuming a seven-year recovery period, what is most nearly the depreciation allowance for the cable in the first year of operation?

(A) \$640

(B) \$670

(C) \$720

(D) \$860

11. A computer with an initial cost of \$1500 and an annual maintenance cost of \$500/yr is purchased and kept indefinitely without any change in its annual maintenance costs. The interest rate is 4%. Most nearly, what is the present worth of all expenditures?

(A) \$12,000

(B) \$13,000

(C) \$14,000

(D) \$15,000

12. A computer with a useful life of 12 years has an initial cost of \$3200 and a salvage value of \$100. The interest rate is 10%. Using the Modified Accelerated Cost Recovery System (MACRS) method of depreciation and a 10-year recovery period, what is most nearly the book value of the computer after the second year?

(A) \$1900

(B) \$2100

(C) \$2300

(D) \$2400

13. A computer with a useful life of 12 years has an initial cost of \$2300 and a salvage value of \$350. The interest rate is 6%. Using the straight-line method, what is most nearly the total depreciation of the computer for the first five years?

(A) $760

(B) $810

(C) $830

(D) $920

14. New 120 mm diameter pipeline is installed over a distance of 5000 m. Annual maintenance and pumping costs are considered to be paid in their entireties at the end of the years in which their costs are incurred. The pipe has the following costs and properties.

initial cost	$2500
annual interest rate	10%
service life	12 yr
salvage value	$300
annual maintenance	$300
pump cost/hour	$1.40
pump operation	600 hr/yr

What is most nearly the equivalent uniform annual cost (EUAC) of the pipe?

(A) $1200

(B) $1300

(C) $1400

(D) $1500

15. A company is considering buying a computer with the following costs and interest rate.

initial cost	$3900
salvage value	$1800
useful life	10 years
annual maintenance	$390
interest rate	6%

Most nearly, what is the equivalent uniform annual cost (EUAC) of the computer?

(A) $740

(B) $780

(C) $820

(D) $850

16. A proposed plant will produce biofuels using fermentation. The plant can be constructed with or without distillation columns to separate the products. The estimated cost and annual net profit with and without the columns are given here.

	plant A (without distillation columns)	plant B (with distillation columns)
cost	$51,000,000	$63,000,000
annual net profit	$6,500,000	$8,200,000

The minimum acceptable rate of return (MARR) is 8% for 15 years. Which of the two plants is a better investment, and what is most nearly the difference in their present worths?

(A) plant A, $9,000,000

(B) plant A, $7,000,000

(C) plant B, $7,000,000

(D) plant B, $2,600,000

17. A company pays off a loan of $1,600,000 by paying the same amount at the end of each year for 20 years. The annual interest rate on the loan is 3.3%. Most nearly, what is the annual payment?

(A) $57,000

(B) $110,000

(C) $320,000

(D) $840,000

18. A power generation station has just acquired a new transformer at a cost of $85,000. Expected maintenance and salvage costs for the first six years are shown.

year	maintenance	salvage
1	$10,000	$40,000
2	$12,000	$38,000
3	$17,000	$35,000
4	$23,000	$25,000
5	$24,000	$24,000
6	$26,000	$22,000

Assuming an interest rate of 10%, after how many years of service should the station replace the transformer?

(A) 2

(B) 3

(C) 4

(D) 5

19. A tank is needed for the foreseeable future. Four options meet the project requirements. Each tank is expected to fail once during its entire lifetime. The cost of such a failure would be $17,700 for any tank. The effective annual interest rate is 10%.

tank	average lifetime (yr)	cost ($)	annual maintenance ($)
1	10	20,300	1010
2	16	28,360	1060
3	25	36,400	1080
4	60	45,080	1130

Which tank should be purchased?

(A) tank 1

(B) tank 2

(C) tank 3

(D) tank 4

20. An engineer is evaluating two options for distributing products from a production facility: renting or owning trucks. The effective annual interest rate is 8%. Most nearly, how many miles must the trucks be driven annually for the two options to have equivalent costs?

rent trucks	own trucks
$0.20/mi	purchase cost of $8000
	useful life of 5 yr
	resale value of $2000 after 5 yr
	annual maintenance cost of $0.09/mi

(A) 8800 mi

(B) 12,000 mi

(C) 15,000 mi

(D) 20,000 mi

Engineering Economics

SOLUTIONS

1. Bring all costs and benefits into the present.

$$\begin{aligned} P_{\text{total}} &= P_{\text{initial}} + P_{\text{maintenance}} - P_{\text{salvage}} \\ &= \$6000 + (\$210)(P/A, 8\%, 5) \\ &\quad -(\$2300)(P/F, 8\%, 5) \\ &= \$6000 + (\$210)(3.9927) \\ &\quad -(\$2300)(0.6806) \\ &= \$5273 \quad (\$5300) \end{aligned}$$

The answer is (B).

2. Add the present worths of all cash flows.

$$\begin{aligned} P_{\text{total}} &= -C_{\text{initial}} - A_{\text{electricity}}(P/A, 10\%, 9) \\ &\quad -A_{\text{maintenance}}(P/A, 10\%, 9) \\ &\quad -A_{\text{insurance}}(P/A, 10\%, 9) \\ &\quad +A_{\text{benefits}}(P/A, 10\%, 9) \\ &\quad +S_9(P/F, 10\%, 9) \\ &= -\$15{,}000 - (\$600)(5.7590) \\ &\quad -(\$120)(5.7590) \\ &\quad -(0.05)(\$15{,}000)(5.7590) + (\$4500)(5.7590) \\ &\quad +(\$900)(0.4241) \\ &= \$2831 \quad (\$2800) \end{aligned}$$

The answer is (B).

3. The effective uniform annual cost (EUAC) is the annual cost equivalent of all costs. When calculating the EUAC, costs are positive and income is negative. Find the annual equivalents of all costs and add them together to get the EUAC.

$$\begin{aligned} \text{EUAC} &= C_{\text{initial}}(A/P, 6\%, 8) \\ &\quad +A_{\text{electricity}} + A_{\text{maintenance}} \\ &\quad +A_{\text{insurance}} - S_8(A/F, 6\%, 8) \\ &= (\$10{,}000)(0.1610) + \$300 + \$100 \\ &\quad +(0.02)(\$10{,}000) \\ &\quad -(\$1000)(0.1010) \\ &= \$2109 \quad (\$2100) \end{aligned}$$

The answer is (B).

4. The annual cost of running and maintaining the 80 mm pipeline is

$$A = \left(\frac{\$2.50}{\text{hr}}\right)(600 \text{ hr}) + \$400 = \$1900$$

Capitalized costs are the present worth of an infinite cash flow.

$$\begin{aligned} P &= \frac{A}{i} = \frac{\$1900}{0.08} \\ &= \$23{,}750 \quad (\$24{,}000) \end{aligned}$$

The answer is (C).

5. The equivalent uniform annual cost (EUAC) is the uniform annual amount equivalent of all cash flows. When calculating the EUAC, costs are positive and income is negative.

$$\begin{aligned} \text{EUAC}_{80} &= A_{\text{initial}} + A_{\text{maintenance}} \\ &\quad +A_{\text{pump}} - A_{\text{salvage}} \\ &= (\$1350)(A/P, 6\%, 6) + \$500 \\ &\quad +\left(\frac{\$2.75}{\text{hr}}\right)(2000 \text{ hr}) \\ &\quad -(\$120)(A/F, 6\%, 6) \\ &= (\$1350)(0.2034) + \$500 \\ &\quad +\$5500 - (\$120)(0.1434) \\ &= \$6257 \quad (\$6300) \end{aligned}$$

The answer is (D).

6. If the unadjusted interest rate is used to calculate the present worth, the answer will be in dollars affected by three years of inflation. To find the uninflated worth three years ago, the effect of inflation during those years must be eliminated from the calculation. To find the answer in uninflated dollars, determine the interest rate adjusted for inflation.

$$\begin{aligned} d &= i + f + (i \times f) \\ &= 0.09 + 0.05 + (0.09)(0.05) \\ &= 0.1445 \end{aligned}$$

Use this adjusted rate in the single payment present worth equation, substituting d for i.

$$\begin{aligned} P &= F(1+d)^{-n} \\ &= (\$5000)(1 + 0.1445)^{-3} \\ &= \$3335 \quad (\$3300) \end{aligned}$$

The answer is (B).

7. The equivalent uniform annual cost (EUAC) is the uniform annual amount equivalent to all cash flows. When calculating the EUAC, costs are positive and income is negative. An expedient way to find the annual worth of the maintenance for the computer is to divide the maintenance costs into two annual series, one of

Engineering Economics

\$275 lasting from year 1 to year 13, and one of \$150 (the difference between \$425 and \$275) lasting from year 9 to year 13. Find the future value in year 13 for each series, add them, and then convert the result back into a single annual amount.

$$\begin{aligned} F_{\$275} &= A(F/A, 6\%, 13) = (\$275)(18.8821) \\ &= \$5192.60 \\ F_{\$150} &= A(F/A, 6\%, 5) = (\$150)(5.6371) \\ &= \$845.60 \\ F_{\text{maintenance}} &= F_{\$275} + F_{\$150} = \$5192.60 + \$845.60 \\ &= \$6038 \\ A_{\text{maintenance}} &= F_{\text{maintenance}}(A/F, 6\%, 13) \\ &= (\$6038)(0.0530) \\ &= \$320 \end{aligned}$$

Calculate the EUAC.

$$\begin{aligned} \text{EUAC} &= A_{\text{initial}} + A_{\text{maintenance}} - A_{\text{salvage}} \\ &= (\$5500)(A/P, 6\%, 13) + \$320 \\ &\quad -(\$3100)(A/F, 6\%, 13) \\ &= (\$5500)(0.1130) + \$320 \\ &\quad -(\$3100)(0.0530) \\ &= \$777 \quad (\$780) \end{aligned}$$

The answer is (B).

8. Use the capitalized cost equation to find the interest rate earned.

$$\begin{aligned} P &= \frac{A}{i} \\ i &= \frac{A_{\text{profit}}}{P_{\text{cost}}} = \frac{\$12{,}000}{\$100{,}000} \\ &= 0.12 \quad (12\%) \end{aligned}$$

The answer is (D).

9. Find the future worth of \$1000.

$$\begin{aligned} F &= P(1+i)^n = (\$1000)(1+0.06)^3 \\ &= \$1191 \quad (\$1190) \end{aligned}$$

The answer is (B).

10. The MACRS factor for the first year, given a seven-year recovery period, is 14.29%.

$$\begin{aligned} D_1 &= (\text{factor})C \\ &= (0.1429)(\$4500) \\ &= \$643 \quad (\$640) \end{aligned}$$

The answer is (A).

11. The expenditures for the computer are the initial cost of \$1500 and the annual maintenance cost of \$500. The annual costs continue indefinitely, so find the present worth of an infinite cash flow.

$$P_{\text{maintenance}} = \frac{A}{i} = \frac{\$500}{0.04} = \$12{,}500$$

The present worth of all expenditures is

$$\begin{aligned} P_{\text{total}} &= P_{\text{initial}} + P_{\text{annual}} = \$1500 + \$12{,}500 \\ &= \$14{,}000 \end{aligned}$$

The answer is (C).

12. Subtract the first two years' depreciation from the original cost.

year	factor (%)	D_j
1	10.00	$(0.10)(\$3200) = \$320$
2	18.00	$(0.18)(\$3200) = \$576$
		$\sum D_j = \$896$

The book value is

$$\begin{aligned} \text{BV} &= \text{initial cost} - \sum D_j = \$3200 - \$896 \\ &= \$2304 \quad (\$2300) \end{aligned}$$

The answer is (C).

13. With the straight-line method, the depreciation is the same every year. Find the annual depreciation.

$$D_j = \frac{C - S_n}{n} = \frac{\$2300 - \$350}{12} = \$162.50$$

The total depreciation for five years is

$$\sum D_{1-5} = (5)(\$162.50) = \$812.50 \quad (\$810)$$

The answer is (B).

Engineering Economics

14. The equivalent uniform annual cost (EUAC) is the uniform annual amount equivalent of all cash flows. When calculating the EUAC, costs are positive and income is negative.

$$\begin{aligned}\text{EUAC}_{120} &= A_{\text{initial}} + A_{\text{maintenance}} \\ &\quad + A_{\text{pump}} - A_{\text{salvage}} \\ &= (\$2500)(A/P, 10\%, 12) + \$300 \\ &\quad + \left(\frac{\$1.40}{\text{hr}}\right)(600\ \text{hr}) \\ &\quad - (\$300)(A/F, 10\%, 12) \\ &= (\$2500)(0.1468) + \$300 + \$840 \\ &\quad - (\$300)(0.0468) \\ &= \$1493 \quad (\$1500)\end{aligned}$$

The answer is (D).

15. The equivalent uniform annual cost (EUAC) is the uniform annual amount equivalent to all cash flows. When calculating the EUAC, costs are positive and income is negative.

$$\begin{aligned}\text{EUAC} &= A_{\text{initial}} + A_{\text{maintenance}} - A_{\text{salvage}} \\ &= (\$3900)(A/P, 6\%, 10) + \$390 \\ &\quad - (\$1800)(A/F, 6\%, 10) \\ &= (\$3900)(0.1359) + \$390 \\ &\quad - (\$1800)(0.0759) \\ &= \$783 \quad (\$780)\end{aligned}$$

The answer is (B).

16. Calculate the total present worth, P_{total}, of each plant. The total present worth consists of the present worth of the annual net profit, P_{net}, minus the present worth of the cost, C.

The present worth of the cost is given. To calculate P_{net}, multiply the annual net profit, A, by the uniform series present worth factor, $(P/A, i\%, n)$. From the tables, $(P/A, 8\%, 15)$ is 8.5595.

For plant A,

$$\begin{aligned}P_{\text{total,A}} &= P_{\text{net,A}} - C_{\text{A}} \\ &= A_{\text{A}}(P/A, 8\%, 15) - C_{\text{A}} \\ &= (\$6{,}500{,}000)(8.5595) - \$51{,}000{,}000 \\ &= \$4{,}636{,}750\end{aligned}$$

For plant B,

$$\begin{aligned}P_{\text{total,B}} &= P_{\text{net,B}} - C_{\text{B}} \\ &= A_{\text{B}}(P/A, 8\%, 15) - C_{\text{B}} \\ &= (\$8{,}200{,}000)(8.5595) - \$63{,}000{,}000 \\ &= \$7{,}187{,}900\end{aligned}$$

$P_{\text{total,B}}$ is greater than $P_{\text{total,A}}$, so plant B is a better investment. The difference in present worth is

$$\begin{aligned}\Delta P &= P_{\text{total,B}} - P_{\text{total,A}} \\ &= \$7{,}187{,}900 - \$4{,}636{,}750 \\ &= \$2{,}551{,}150 \quad (\$2{,}600{,}000)\end{aligned}$$

The answer is (D).

17. Use the formula for capital recovery. The present worth, P, is the amount of the loan. The annual payment for the loan is

$$\begin{aligned}A &= P\left(\frac{i(1+i)^n}{(1+i)^n - 1}\right) \\ &= (\$1{,}600{,}000)\left(\frac{(0.033)(1+0.033)^{20}}{(1+0.033)^{20} - 1}\right) \\ &= \$110{,}550 \quad (\$110{,}000)\end{aligned}$$

The answer is (B).

18. Use the equivalent uniform annual cost (EUAC) as a basis for comparison. Calculate the EUAC of the transformer for one year of service, two years, three years, and so on. Continue calculating as long as the EUAC decreases for each year added. When the EUAC for $n+1$ years is greater than the EUAC for n years, stop calculating; the transformer should be replaced after n years.

To calculate the EUAC, convert the purchase price, future maintenance costs, and future salvage value to equivalent annual values and add them.

For one year,

$$\begin{aligned}\text{EUAC}_1 &= A_{\text{cost}} + A_{\text{maint}} + A_{\text{salvage}} \\ &= P_{\text{cost}}(A/P, 10\%, 1) + F_{\text{maint}}(A/F, 10\%, 1) \\ &\quad - F_{\text{salvage}}(A/F, 10\%, 1) \\ &= (\$85{,}000)(1.1000) + (\$10{,}000)(1.0000) \\ &\quad - (\$40{,}000)(1.0000) \\ &= \$63{,}500\end{aligned}$$

To combine varying maintenance costs for different years into a single annual value, calculate the present worth of each year's maintenance cost, add the present worths together, and convert the total to an equivalent annual value.

For two years, the present value of the maintenance costs is

$$\begin{aligned} P_{\text{maint}} &= F_{\text{maint,yr 1}}(P/F, 10\%, 1) \\ &\quad + F_{\text{maint,yr 2}}(P/F, 10\%, 2) \\ &= (\$10{,}000)(0.9091) + (\$12{,}000)(0.8264) \\ &= \$19{,}008 \end{aligned}$$

The EUAC for two years of service is

$$\begin{aligned} \text{EUAC}_2 &= A_{\text{cost}} + A_{\text{maint}} + A_{\text{salvage}} \\ &= P_{\text{cost}}(A/P, 10\%, 2) + P_{\text{maint}}(A/P, 10\%, 2) \\ &\quad - F_{\text{salvage}}(A/F, 10\%, 2) \\ &= (\$85{,}000)(0.5762) + (\$19{,}007.80)(0.5762) \\ &\quad - (\$38{,}000)(0.4762) \\ &= \$41{,}834 \end{aligned}$$

EUAC_2 is less than EUAC_1, so continue. For three years, the present value of the maintenance costs is

$$\begin{aligned} P_{\text{maint}} &= F_{\text{maint,yr 1}}(P/F, 10\%, 1) \\ &\quad + F_{\text{maint,yr 2}}(P/F, 10\%, 2) \\ &\quad + F_{\text{maint,yr 3}}(P/F, 10\%, 3) \\ &= (\$10{,}000)(0.9091) + (\$12{,}000)(0.8264) \\ &\quad + (\$17{,}000)(0.7513) \\ &= \$31{,}780 \end{aligned}$$

The EUAC for three years of service is

$$\begin{aligned} \text{EUAC}_3 &= P_{\text{cost}}(A/P, 10\%, 3) + P_{\text{maint}}(A/P, 10\%, 3) \\ &\quad - F_{\text{salvage}}(A/F, 10\%, 3) \\ &= (\$85{,}000)(0.4021) + (\$31{,}779.90)(0.4021) \\ &\quad - (\$35{,}000)(0.3021) \\ &= \$36{,}384 \end{aligned}$$

EUAC_3 is less than EUAC_2, so continue. For four years, the present value of the maintenance costs is

$$\begin{aligned} P_{\text{maint}} &= F_{\text{maint,yr 1}}(P/F, 10\%, 1) \\ &\quad + F_{\text{maint,yr 2}}(P/F, 10\%, 2) \\ &\quad + F_{\text{maint,yr 3}}(P/F, 10\%, 3) \\ &\quad + F_{\text{maint,yr 4}}(P/F, 10\%, 4) \\ &= (\$10{,}000)(0.9091) + (\$12{,}000)(0.8264) \\ &\quad + (\$17{,}000)(0.7513) + (\$23{,}000)(0.6830) \\ &= \$47{,}489 \end{aligned}$$

The EUAC for four years of service is

$$\begin{aligned} \text{EUAC}_4 &= P_{\text{cost}}(A/P, 10\%, 4) + P_{\text{maint}}(A/P, 10\%, 4) \\ &\quad - F_{\text{salvage}}(A/F, 10\%, 4) \\ &= (\$85{,}000)(0.3155) + (\$47{,}488.90)(0.3155) \\ &\quad - (\$25{,}000)(0.2155) \\ &= \$36{,}413 \end{aligned}$$

EUAC_4 is greater than EUAC_3, so the transformer should be replaced after three years of service.

The answer is (B).

19. The four options have different lifetimes, so it is necessary to compare them on an equivalent uniform annual cost (EUAC) basis. The annual cost sums the annualized initial investment, the annual maintenance cost, and the cost of failure multiplied by the probability of failure.

$$\begin{aligned} \text{EUAC} &= P(A/P, 10\%, n) + M + \frac{1}{n}F \\ \text{EUAC}_1 &= (\$20{,}300)(0.1627) + \$1010 \\ &\quad + \left(\frac{1}{10\ \text{yr}}\right)(\$17{,}700) \\ &= \$6083/\text{yr} \\ \text{EUAC}_2 &= (\$28{,}360)(0.1278) + \$1060 \\ &\quad + \left(\frac{1}{16\ \text{yr}}\right)(\$17{,}700) \\ &= \$5791/\text{yr} \\ \text{EUAC}_3 &= (\$36{,}400)(0.1102) + \$1080 \\ &\quad + \left(\frac{1}{25\ \text{yr}}\right)(\$17{,}000) \\ &= \$5799/\text{yr} \\ \text{EUAC}_4 &= (\$45{,}080)(0.1003) + \$1130 \\ &\quad + \left(\frac{1}{60\ \text{yr}}\right)(\$17{,}700) \\ &= \$5947/\text{yr} \end{aligned}$$

It's not the cards you're dealt in life

It's how you play them

Tank 2 has the lowest annualized cost.

The answer is (B).

20. Let x represent the number of miles are driven annually. The costs for the two options are

$$\text{cost}_{\text{rent}} = \left(\$0.20\ \frac{\$}{\text{mi}}\right)x$$

$$\begin{aligned}\text{cost}_{\text{own}} &= \left(\$0.09\ \frac{\$}{\text{mi}}\right)x + (\$8000)(A/P, 8\%, 5) \\ &\quad -(\$2000)(A/F, 8\%, 5) \\ &= \left(\$0.09\ \frac{\$}{\text{mi}}\right)x + (\$8000)(0.2505) \\ &\quad -(\$2000)(0.1705)\end{aligned}$$

Calculate the number of miles driven annually when the cost of the two options is equal.

$$\begin{aligned}\text{cost}_{\text{rent}} &= \text{cost}_{\text{own}} \\ \left(\$0.20\ \frac{\$}{\text{mi}}\right)x &= \left(\$0.09\ \frac{\$}{\text{mi}}\right)x \\ &\quad +(\$8000)(0.2505) \\ &\quad -(\$2000)(0.1705) \\ \left(\$0.20\ \frac{\$}{\text{mi}}\right)x - \left(\$0.09\ \frac{\$}{\text{mi}}\right)x &= \$1663 \\ \left(\$0.11\ \frac{\$}{\text{mi}}\right)x &= \$1663 \\ x &= \frac{\$1663}{\$0.11\ \frac{\$}{\text{mi}}} \\ &= 15{,}118\ \text{mi} \quad (15{,}000\ \text{mi})\end{aligned}$$

The answer is (C).

Engineering Economics

43 Professional Practice

PRACTICE PROBLEMS

1. What must be proven for damages to be collected from a strict liability in tort?

(A) that willful negligence caused an injury

(B) that willful or unwillful negligence caused an injury

(C) that the manufacturer knew about a product defect before the product was released

(D) none of the above

2. A material breach of contract occurs when the

(A) contractor uses material not approved by the contract for use

(B) contractor's material order arrives late

(C) owner becomes insolvent

(D) contractor installs a feature incorrectly

3. If a contract has a value engineering clause and a contractor suggests to the owner that a feature or method be used to reduce the annual maintenance cost of the finished project, what will be the most likely outcome?

(A) The contractor will be able to share one time in the owner's expected cost savings.

(B) The contractor will be paid a fixed amount (specified by the contract) for making a suggestion, but only if the suggestion is accepted.

(C) The contract amount will be increased by some amount specified in the contract.

(D) The contractor will receive an annuity payment over some time period specified in the contract.

4. A tort is

(A) a civil wrong committed against another person

(B) a section of a legal contract

(C) a legal procedure in which complaints are heard in front of an arbitrator rather than a judge or jury

(D) the breach of a contract

5. If a contract does not include the boilerplate clause, "Time is of the essence," which of the following is true?

(A) It is difficult to recover losses for extra hours billed.

(B) Standard industry time guidelines apply.

(C) Damages for delay cannot be claimed.

(D) Workers need not be paid for downtime in the project.

6. Which statement is true regarding the legality and enforceability of contracts?

(A) For a contract to be enforceable, it must be in writing.

(B) A contract to perform illegal activity will still be enforced by a court.

(C) A contract must include a purchase order.

(D) Mutual agreement of all parties must be evident.

7. Which option best describes the contractual lines of privity between parties in a general construction contract?

(A) The consulting engineer will have a contractual obligation to the owner, but will not have a contractual obligation with the general contractor or the subcontractors.

(B) The consulting engineer will have a contractual obligation to the owner and the general contractor.

(C) The consulting engineer will have a contractual obligation to the owner, general contractor, and subcontractors.

(D) The consulting engineer will have a contractual obligation to the general contractor, but will not have a contractual obligation to the owner or subcontractors.

8. A contract has a value engineering clause that allows the parties to share in improvements that reduce cost. The contractor had originally planned to transport concrete on-site for a small pour with motorized wheelbarrows. On the day of the pour, however, a concrete pump is available and is used, substantially reducing the contractor's labor cost for the day. This is an example of

(A) value engineering whose benefit will be shared by both contractor and owner

(B) efficient methodology whose benefit is to the contractor only

(C) value engineering whose benefit is to the owner only

(D) cost reduction whose benefit will be shared by both contractor and laborers

9. In which of the following fee structures is a specific sum paid to the engineer for each day spent on the project?

(A) salary plus

(B) per-diem fee

(C) lump-sum fee

(D) cost plus fixed fee

10. What type of damages is paid when responsibility is proven but the injury is slight or insignificant?

(A) nominal

(B) liquidated

(C) compensatory

(D) exemplary

11. One of the main reasons to enter into a joint venture on a big construction project is to

(A) let bidders know the size of the project for more accurate cost estimating

(B) better plan future uses for the completed project

(C) spread the risks associated with the project

(D) pay for the bidding process

SOLUTIONS

1. In order to prove strict liability in tort, it must be shown that a product defect caused an injury. Negligence need not be proven, nor must the manufacturer know about the defect before release.

The answer is (D).

2. A material breach of the contract is a significant event that is grounds for cancelling the contract entirely. Typical triggering events include failure of the owner to make payments, the owner causing delays, the owner declaring bankruptcy, the contractor abandoning the job, or the contractor getting substantially off schedule.

The answer is (C).

3. Changes to a structure's performance, safety, appearance, or maintenance that benefit the owner in the long run will be covered by the value engineering clause of a contract. Normally, the contractor is able to share in cost savings in some manner by receiving a payment or credit to the contract.

The answer is (A).

4. A tort is a civil wrong committed against a person or his/her property which results in some form of damages. Torts are normally resolved through lawsuits.

The answer is (A).

5. This clause must be included in order to recover damages due to delay.

The answer is (C).

6. In order for a contract to be legally binding, it must

- be established for a legal purpose
- contain a mutual agreement by all parties
- have consideration, or an exchange of something of value (e.g., a service is provided in exchange for a fee)
- not obligate parties to perform illegal activity
- not be between parties that are mentally incompetent, minors, or do not otherwise have the power to enter into the contract

A contract does not need to use as its basis or include a purchase order to be enforceable. Oral contracts may be legally binding in some instances, depending on the circumstances and purpose of the contract. Oral contracts may be difficult to enforce, however, and should not be used for engineering and construction agreements.

The answer is (D).

7. With a general construction contract, a consulting engineer will be hired by the owner to develop the design and contract documents, as well as to assist in the preparation of the bid documents and provide contract administrative services during the construction phase. The contract documents produced by the engineer will form the basis of the owner's agreement with the contractor. Although the engineer will work closely with the contractor during the construction phase, and may work with subcontractors as well, the engineer will not have a contractual line of privity with either party.

The answer is (A).

8. The problem gives an example of efficient methodology, where the benefit is to the contractor only. It is not an example of value engineering, as the change affects the contractor, not the owner. Performance, safety, appearance, and maintenance are unaffected.

The answer is (B).

9. A specific fee is paid to the engineer for each day on the job in a per-diem fee structure.

The answer is (B).

10. Nominal damages are awarded for inconsequential injuries.

The answer is (A).

11. Risk is an important factor in construction. Therefore, it is prudent to spread it as widely as possible. One way to do this is to enter into a joint venture with other contractors. A joint venture is a short-term partnership arrangement in which each of two or more participating construction companies is committed to a predetermined percentage of a contract, and each shares proportionately in the final profit or loss. One of the participating companies acts as the manager or sponsor of the project.

The answer is (C).

44 Ethics

PRACTICE PROBLEMS

1. An environmental engineer with five years of experience reads a story in the daily paper about a proposal being presented to the city council to construct a new sewage treatment plant near protected wetlands. Based on professional experience and the facts presented in the newspaper, the engineer suspects the plant would be extremely harmful to the local ecosystem. Which of the following would be an acceptable course of action?

(A) The engineer should contact appropriate agencies to get more data on the project before making a judgment.

(B) The engineer should write an article for the paper's editorial page urging the council not to pass the project.

(C) The engineer should circulate a petition through the community condemning the project, and present the petition to the council.

(D) The engineer should do nothing because he doesn't have enough experience in the industry to express a public opinion on the matter.

2. An engineer is consulting for a construction company that has been receiving bad publicity in the local papers about its waste-handling practices. Knowing that this criticism is based on public misperceptions and the paper's thirst for controversial stories, the engineer would like to write an article to be printed in the paper's editorial page. What statement best describes the engineer's ethical obligations?

(A) The engineer's relationship with the company makes it unethical for him to take any public action on its behalf.

(B) The engineer should request that a local representative of the engineering registration board review the data and write the article in order that an impartial point of view be presented.

(C) As long as the article is objective and truthful, and presents all relevant information including the engineer's professional credentials, ethical obligations have been satisfied.

(D) The article must be objective and truthful, present all relevant information including the engineer's professional credentials, and disclose all details of the engineer's affiliation with the company.

3. After making a presentation for an international project, an engineer is told by a foreign official that his company will be awarded the contract, but only if it hires the official's brother as an advisor to the project. The engineer sees this as a form of extortion and informs his boss. His boss tells him that, while it might be illegal in the United States, it is a customary and legal business practice in the foreign country. The boss impresses upon the engineer the importance of getting the project, but leaves the details up to the engineer. What should the engineer do?

(A) He should hire the official's brother, but insist that he perform some useful function for his salary.

(B) He should check with other companies doing business in the country in question, and if they routinely hire relatives of government officials to secure work, then he should do so too.

(C) He should withdraw his company from consideration for the project.

(D) He should inform the government official that his company will not hire the official's brother as a precondition for being awarded the contract, but invite the brother to submit an application for employment with the company.

4. If one is aware that a registered engineer willfully violates a state's rule of professional conduct, one should

(A) do nothing

(B) report the violation to the state's engineering registration board

(C) report the violation to the employer

(D) report the violation to the parties it affects

5. Which of the following is an ethics violation specifically included in the National Council of Examiners for Engineering and Surveying (NCEES) *Model Rules*?

(A) an engineering professor "moonlighting" as a private contractor

(B) an engineer investing money in the stock of the company for which he/she works

(C) a civil engineer with little electrical experience signing the plans for an electric generator

(D) none of the above

6. A senior licensed professional engineer with 30 years of experience in geotechnical engineering is placed in charge of a multidisciplinary design team consisting of a structural group, a geotechnical group, and an environmental group. In this role, the engineer is responsible for supervising and coordinating the efforts of the groups when working on large interconnected projects. In order to facilitate coordination, designs are prepared by the groups under the direct supervision of the group leader, and then they are submitted to her for review and approval. This arrangement is ethical as long as

(A) the engineer signs and seals each design segment only after being fully briefed by the appropriate group leader

(B) the engineer signs and seals only those design segments pertaining to geotechnical engineering

(C) each design segment is signed and sealed by the licensed group leader responsible for its preparation

(D) the engineer signs and seals each design segment only after it has been reviewed by an independent consulting engineer who specializes in the field in which it pertains

7. The National Society of Professional Engineers' (NSPE) *Code of Ethics for Engineers* addresses competitive bidding. Which of the following is NOT stipulated?

(A) Engineers and their firms may refuse to bid competitively on engineering services.

(B) Clients are required to seek competitive bids for design services.

(C) Federal laws governing procedures for procuring engineering services (e.g., competitive bidding) remain in full force.

(D) Engineers and their societies may actively lobby for legislation that would prohibit competitive bidding for design services.

8. A city engineer is in charge of receiving bids on behalf of the city council. A contractor's bid arrives with two tickets to a professional football game. The bid is the lowest received. What should the engineer do?

(A) Return the tickets and accept the bid.

(B) Return the tickets and reject the bid.

(C) Discard the tickets and accept the bid.

(D) Discard the tickets and reject the bid.

9. A relatively new engineering firm is considering running an advertisement for their services in the local newspaper. An ad agency has supplied them with four concepts. Of the four ad concepts, which one(s) would be acceptable from the standpoint of professional ethics?

I. an advertisement contrasting their successes over the past year with their nearest competitors' failures

II. an advertisement offering a free television to anyone who hires them for work valued at over $10,000

III. an advertisement offering to beat the price of any other engineering firm for the same services

IV. an advertisement that tastefully depicts their logo against the backdrop of the Golden Gate Bridge

(A) I and III

(B) I, III, and IV

(C) II, III, and IV

(D) neither I, II, III, nor IV

10. A professional engineer who took the licensing examination in mechanical engineering may

(A) not design in electrical engineering

(B) design in electrical engineering if she feels competent

(C) design in electrical engineering if she feels competent and the electrical portion of the design is insignificant and incidental to the overall job

(D) design in electrical engineering if another engineer checks the electrical engineering work

11. An engineering firm is hired by a developer to prepare plans for a shopping mall. Prior to the final bid date, several contractors who have received bid documents and plans contact the engineering firm with requests for information relating to the project. What can the engineering firm do?

(A) The firm can supply requested information to the contractors as long as it does so fairly and evenly. It cannot favor or discriminate against any contractor.

(B) The firm should supply information to only those contractors that it feels could safely and economically perform the construction services.

(C) The firm cannot reveal facts, data, or information relating to the project that might prejudice a contractor against submitting a bid on the project.

(D) The firm cannot reveal facts, data, or information relating to the project without the consent of the client as authorized or required by law.

12. Which of the following is/are unethical for an engineer to accept from a supplier?

(A) special discounts

(B) gifts

(C) commissions

(D) all of the above

13. Which of the following forms of advertising is ethical for a contractor?

(A) television and radio ads

(B) flyers on houses with special offers

(C) asking clients to tell their friends

(D) all of the above

14. The National Council of Examiners for Engineering and Surveying (NCEES) *Model Rules* allow a registered engineer to express publicly a professional opinion under which of the following circumstances?

(A) whenever the engineer's opinion is requested by a supervisor

(B) whenever the engineer cares to freely give such opinions

(C) whenever the engineer's opinion is based on competent evaluation of the subject matter and adequate knowledge of the relevant facts

(D) after the engineer has been professionally employed for a minimum length of time

SOLUTIONS

1. The engineer certainly has more experience and knowledge in the field than the general public or even the council members who will have to vote on the issue. Therefore, the engineer is qualified to express his opinion if he wishes to do so. Before the engineer takes any public position, however, the engineer is obligated to make sure that all the available information has been collected.

The answer is (A).

2. It is ethical for the engineer to issue a public statement concerning a company he works for, provided he makes that relationship clear and provided the statement is truthful and objective.

The answer is (D).

3. Hiring the official's brother as a precondition for being awarded the contract is a form of extortion. Depending on the circumstances, however, it may be legal to do so according to U.S. law. (The Foreign Corrupt Practices Act of 1977 allows American companies to pay extortion in some cases.) This practice, however, is not approved by the National Council of Examiners for Engineering and Surveying (NCEES) *Model Rules*:

> Registrants shall not offer, give, solicit, or receive, either directly or indirectly, any commission or gift, or other valuable consideration in order to secure work.

The answer is (D).

4. A violation should be reported to the organization that has promulgated the rule.

The answer is (B).

5. The National Council of Examiners for Engineering and Surveying (NCEES) *Model Rules* specifically states that registrants may not perform work beyond their level of expertise. The other two examples may be unethical under some circumstances, but are not specifically forbidden by the NCEES code.

The answer is (C).

6. According to the National Council of Examiners for Engineering and Surveying (NCEES) *Model Rules*,

> Licensees may accept assignments for coordination of an entire project, provided that each design segment is signed and sealed by the registrant responsible for preparation of that design segment.

The answer is (C).

7. Clients are not required to seek competitive bids. In fact, many engineering societies discourage the use of bidding to procure design services because it is believed that competitive bidding results in lower-quality construction.

The answer is (B).

8. Registrants should not accept gifts from parties expecting special consideration, so the tickets cannot be kept. They also should not be merely discarded, for several reasons. Inasmuch as the motive of the contractor is not known with certainty, in the absence of other bidding rules, the bid may be accepted.

The answer is (A).

9. None of the ads is acceptable from the standpoint of professional ethics. Concepts I and II are explicitly prohibited by the National Council of Examiners for Engineering and Surveying (NCEES) *Model Rules*. Concept III demeans the profession of engineering by placing the emphasis on price as opposed to the quality of services. Concept IV is a misrepresentation; the picture of the Golden Gate Bridge in the background might lead some potential clients to believe that the engineering firm in question had some role in the design or construction of that project.

The answer is (D).

10. Although the laws vary from state to state, engineers are usually licensed generically. Engineers are licensed as "professional engineers." The scope of their work is limited only by their competence. In the states where the license is in a particular engineering discipline, an engineer may "touch upon" another discipline when the work is insignificant and/or incidental.

The answer is (C).

11. It is normal for engineers and architects to clarify the bid documents. However, some information may be proprietary to the developer. The engineering firm should only reveal information that has already been publicly disseminated or approved for release with the consent of the client.

The answer is (D).

12. Engineers may not solicit or accept any special treatment from a supplier. This is to ensure fair competition among suppliers.

The answer is (D).

13. Most codes of ethics state that engineers should build their reputations based upon the merits of their work. Some forms of print advertising are acceptable, but advertisements must be done in a tasteful manner so as not to damage the dignity of the profession. Also, contractors are often required to list their license number on any form of advertising.

The answer is (C).

14. Unless a registered engineer has a competent evaluation of the matter, as well as knowledge of all the relevant facts, the NCEES *Model Rules* do not allow an engineer to express their professional opinion publicly.

The answer is (C).

45 Licensure

PRACTICE PROBLEMS

1. Which organization provides the authority for the private practice of engineering?

(A) the National Council of Examiners for Engineering and Surveying (NCEES)

(B) the Accreditation Board for Engineering and Technology (ABET)

(C) the United States Department of Commerce

(D) the state in which an engineer lives

2. Reciprocity is a term that describes the process by which one state may honor an engineer's professional engineering license from another state. Which word is commonly used as a synonym for "reciprocity"?

(A) normality

(B) comity

(C) suplurity

(D) anority

3. What term best defines the reason that engineers working for some companies do not need to be licensed as professional engineers?

(A) commercial exclusion

(B) corporate oversight

(C) industrial exemption

(D) caveat emptor

4. The purpose of professional licensure of engineers is to

(A) protect the engineering profession

(B) ensure public health

(C) limit competition to licensed engineers in a state

(D) protect engineering titles

SOLUTIONS

1. State laws provide the authority for professional practice. NCEES writes the examinations used by the states. ABET accredits four-year degree programs. The United States Department of Commerce is not involved.

The answer is (D).

2. The terms comity and reciprocity, though slightly different in meaning, are often used synonymously.

The answer is (B).

3. Professional licensure is intended to protect the public from engineers in private practice. States have various laws, commonly referred to as "industrial exemptions," that allow engineers to work in the industry (e.g., in companies that produce products) without being licensed.

The answer is (C).

4. Although engineering licensure laws do protect engineering titles, the purpose of professional licensure is to protect public health, safety, and welfare from unqualified practitioners.

The answer is (B).